THE FIRST INTO THE DARK

Michael Robertson
Astrid Ley
Edwina Light

THE FIRST INTO THE DARK

THE NAZI PERSECUTION OF THE DISABLED

UTS ePRESS
University of Technology Sydney
Broadway NSW 2007 AUSTRALIA
epress.lib.uts.edu.au

Copyright Information
This book is copyright. The work is licensed under a Creative Commons Attribution Non Commercial-Non Derivatives License CC BY-NC-ND

http://creativecommons.org/licenses/by-nc-nd/4.0

First Published 2019
© 2019 in the text and images, the authors
© 2019 in the book design and layout, UTS ePRESS

Publication Details
DOI citation: https://doi.org/10.5130/aae
ISBN: 978-0-6481242-2-1 (paperback)
ISBN: 978-0-6481242-3-8 (pdf)
ISBN: 978-0-6481242-5-2 (ePub)
ISBN: 978-0-6481242-6-9 (Mobi)

Peer Review
This work was peer reviewed by disciplinary experts.

Declaration of conflicting interest
The editors declare no potential conflicts of interest with respect to the research, authorship, and/or publication of this book.

Funding
The editors received no dedicated financial support for the research and publication of this book.

UTS ePRESS
Manager: Scott Abbott
Books Editor: Matthew Noble
Designer: Emily Gregory
Enquiries: utsepress@uts.edu.au

For enquiries about third party copyright material reproduced in this work, please contact UTS ePRESS.

OPEN ACCESS
UTS ePRESS publishes peer reviewed books, journals and conference proceedings and is the leading publisher of peer reviewed open access journals in Australasia. All UTS ePRESS online content is free to access and read.

To read the free, open access version of this book online, visit https://doi.org/10.5130/aae or scan this QR code with your mobile device:

EPIGRAPH

DOG FOX FIELD

The test for feeblemindedness was, they had to make up a sentence using the words dog, fox *and* field.
Judgement at Nuremberg

These were no leaders, but they were first
into the dark on Dog Fox Field:

Anna who rocked her head, and Paul
who grew big and yet giggled small,

Irma who looked Chinese, and Hans
who knew his world as a fox knows a field.

Hunted with needles, exposed, unfed,
this time in their thousands they bore sad cuts

for having gaped, and shuffled, and failed
to field the lore of prey and hound

they then had to thump and cry in the vans
that ran while stopped in Dog Fox Field.

Our sentries, whose holocaust does not end,
they show us when we cross into Dog Fox Field.

- Les Murray, from the book *Dog fox field*.
By kind permission

TABLE OF CONTENTS

Epigraph v
Table of contents vii
List of illustrations ix
Acknowledgements xv

Part 1 : The Crime **3**
    Part 1 - Introduction 5
    Chapter 1 : The iron door 9
    Chapter 2 : 'Life unworthy of living' 25
    Chapter 3 : *Aktion* T4 – Implementation and resistance 39
    Chapter 4 : 'Dr Schneider' 65
    Chapter 5 : 'There was wonderful material among those brains' 93
    Chapter 6 : 'Cure or destroy' 117

Part 2 : Lessons and legacies **139**
    Part 2 - Introduction 141
    Chapter 7 : Was Elvira Hempel a survivor of the Holocaust? 145
    Chapter 8 : The ethical dimension 165
    Chapter 9 : Remembrance, commemoration, memorialisation 209
    Epilogue 233

Endnotes 247
Glossary 301

Appendix 1 – Key historical figures and institutions 305
Appendix 2 – Other crimes perpetrated by non-German scientists in the mid-twentieth century 311
Appendix 3 – Note on research and sources 315
Appendix 4 – Select English language bibliography 319

Author biographies 334

LIST OF ILLUSTRATIONS

FIGURE

1. The Uchtspringe Institution in 1903
    Uchtspringe-1903-Heil-und Pflegeanstalt (Healing and Nursing Home). Creative Commons, Public domain

2. The Uchtspringe Institution in the 1940s
    Brandenburg Memorials Foundation

3. Notes from the assessment of Elvira Hempel by Professor Fünfgeld in 1938
    Photo courtesy of Heinz Manthey

4. Elvira Hempel (left) and Lisa Hempel (right)
    Photo courtesy of Heinz Manthey

5. The Brandenburg extermination centre in the 1940s
    Photo courtesy of Brandenburg Memorials Foundation

6. Direction from Adolf Hitler to Philipp Bouhler and Karl Brandt, 1 September 1939
    Dokumentationszentrum Reichsparteitagsgelände. Public domain

7. Deborah Kallikak at work in the sewing room of the Vineland Training School for Feebleminded Children, c1910
    Wikimedia Commons: The Kallikak Family. Public domain

8. A registration card for the *Aktion* T4 program
   Patientenarchiv im Christophsbad, Göppingen, Germany
9. Patients at the 'Schloß Bruckberg' care institution
   Pressestelle der Diakonie Neuendettelsau. Public domain
10. The *Aktion* T4 killing centres in Nazi Germany and Austria 1939-1941 in their present day locations
    Authors illustration. Creative Commons CC-BY-NC-ND
11. Anti-disability propaganda image published in the journal *Neues Volk*
    Photo: From exhibition 'Die Landesanstalt Görden 1933 bis 1945: Psychiatrie im Nationalsozialismus', Görden/Brandenburg. Reproduced with Permission. Author photo, November 2016
12. Archbishop of Münster Clemens von Galen
    Photo: Diocesan Archives, Münster. Reproduced with Permission
13. Karl Bonhoeffer, professor of psychiatry at Charité Hospital in Berlin
    Photo: Monatsschrift für Psychiatrie und Neurologie 1938. Public Domain
14. The gas chamber at Hadamar 'euthanasia' memorial (present day)
    Hadamar Gedenkstätte. Author photo, July 2015
15. Installation from the exhibition at Psychiatrie Museum kbo-Isar-Amper-Klinikum (Eglfing-Haar) depicting the portions of starvation rations
    Author photo, July 2013
16. Irmfried Eberl (1942)
    Photo: State Archives, Ludwigsburg
17. The Theresienstadt 'camp-ghetto' site (present day)
    Author photo, July 2017
18. *Einsatzgruppen* murdering a group of Jewish victims in Ivanagorod in Ukraine, 1942
    Photo: Zwiazek Bojowników o Wolnosc i Demokracje (1959). Public Domain

List of Illustrations

19. A Gaswagen (gas van or mobile gas chamber)
    Photo: Office of the United States Chief Counsel for Prosecution of Axis Criminality publication Nazi Conspiracy and Aggression; Washington, U.S Govt. Print. Office, 1946, Vol III, p. 418. Public Domain

20. Map of the Treblinka extermination camp
    Wikimedia Commons: Bundesarchiv, picture 183-F0918-0201-001 / Creative Commons CC-BY-SA 3.0

21. Werner Przadaka, a 13-year-old boy murdered on 28 October 1940 at the Brandenburg killing centre
    Photo: Federal Archives, Berlin. Reproduced with Permission

22. Julius Hallervorden
    Photo: Archives of Max-Planck Society.
    Reproduced with Permission

23. The developmental assessment conducted on Elvira Manthey
    Photo courtesy of Heinz Manthey

24. Hans Heinze, Director of the Brandenburg-Görden State Hospital 1938-1945
    Photo: From exhibition 'Die Landesanstalt Görden 1933 bis 1945: Psychiatrie im Nationalsozialismus', Görden/Brandenburg. Reproduced with Permission. Author photo, November 2016

25. A dissection table at the Hadamar killing centre (present day)
    Hadamar Gedenkstätte. Author photo, July 2015

26. Leo Alexander assisting the prosecution at the Nuremberg Doctors' trial 1947-1948
    Photo: National Archives and Records Administration, College Park Source Record ID: 111-SC-275098 (Album 5576). Public Domain

27. A prototype Electroconvulsive Therapy (ECT) machine
    Photo: On display at the Psychiatrie Museum kbo-Isar-Amper-Klinikum (Eglfing-Haar), München-Ost. Author photo, July 2013

28. Herman Paul Nitsche
    Photo: www.HolocaustResearchProject.org.
    Creative Commons CC0

29. The former Heilanstalt at Pirna-Sonnenstein (present day)
    Author photo, July 2015

30. Georg Mall
    Photo: From Exhibition MuSeele Christophsbad Göppingen.
    Author photo, July 2015

31. Pages from a post war text book produced by psychiatrist Gerhart Mall
    Photo: From Exhibition MuSeele Christophsbad Göppingen.
    Author photo, July 2015

32. The Dresden 'Doctors' trial', 1947-1948
    Photo: Bundesarchiv, Bild 183-H26186.
    Creative Commons CC-BY-SA 3.0

33. The historical relationship between the *Krankenmorde* and the Shoah
    Authors illustration. Creative Commons CC-BY-NC-ND

34. Nazi anti-tobacco propaganda from 1941
    Wikimedia Commons: File:German_anti-smoking

35. Karl Brandt at the Nuremburg Doctors' trial, 1947-1948
    Photo: 20 August 1947. From Public Relations Photo Section, Office Chief of Counsel for War Crimes, Nuremberg, Germany, APO 696-A, US Army. Photo No. OMT-I-D-144. Citation: Telford Taylor Papers, Arthur W. Diamond Law Library, Columbia University Law School, New York, N.Y.: TTP-CLS: 15-1-1-76. Public Domain

36. Disability Rights poster commissioned by the World Health Organisation
    From: http://www.who.int/disabilities/world_report/2011/posters/en/

List of Illustrations

37. The post mortem report for Babette Fröwis dated November 1943
    Document image: Archiv des Bezirk Oberbayern.
    Reproduced with Permission

38. The Memorial for victims of the Nazi 'Euthanasia' Program on the original site of the Tiergartenstraße 4 villa, Berlin
    Author photo, July 2015

39. Trees on the hillside below the Pirna-Sonnenstein memorial site
    Author photo, July 2015

40. Frank Schneider delivers the formal apology of the German Psychiatric Association to victims of the *Krankenmorde* and their families in Berlin on 26 November 2010
    Photo courtesy of Frank Schneider

41. Hans and Elvira Manthey on their wedding day in September 1966
    Photo courtesy of Heinz Manthey

42. Elvira Manthey (front) visits the Uchtspringe Institution in the early 1990s
    Photo courtesy of Heinz Manthey

43. Elvira Manthey visits the memorial on the site of the Brandenburg killing centre (1992)
    Photo courtesy of Heinz Manthey

ACKNOWLEDGEMENTS

This project began in 2013 when the three authors met at the memorial site at Brandenburg-an-der-Havel during a period of research field work in Germany and Austria. The remarkable story of Elvira Hempel's survival of the Nazi's 'euthanasia' program and the many questions it raised prompted us to collaborate over the next five years in a program of work that led to the creation of this book. Creating a work worthy of the story of the many victims of the Nazi regime's persecution of its citizens living with illness and disability required not only time, but drew on the generous collegiality of others. We shared innumerable hours visiting multiple archives, memorial sites and institutions in Germany, Austria and Poland, and interacted with many historians, archivists, museum curators, educators and other experts in the field. As any Holocaust scholar would attest, to study a crime of such magnitude requires a sense of the landscapes involved—to stand at the railhead of Auschwitz-Birkenau in Poland or in the killing field at the Ponary forest in Lithuania is pre-requisite to even begin to know such events. By analogy, the same must be said of the Nazi 'euthanasia' program. We could not have begun to write this work without, for example, walking into the gas chamber site at the Hadamar memorial or the courtrooms at the *Justizpalast* in Nuremberg where perpetrators were tried and sentenced for their crimes.

The next challenge was to construct our work by uniting three voices from multiple disciplinary perspectives—psychiatry, mental health ethics, history, bioethics, and journalism. With the signal intent of narrating the story of Elvira Hempel and many other victims of the Nazi persecution of the sick and disabled, as well as gradually revealing those others who played a role as persecutor, bystander, protector or liberator, we hoped to bring these events into contemporary consciousness. We sought to achieve this through a sober consideration of what this historical moment exposes then and now through dilemmas related to medical ethics, human rights, and decisions about the beginning and end of life. With the editorial advice of Matthew Noble and colleagues at UTS ePRESS, we were able to refine *The First into the Dark* into a story that is both accessible to all readers and more worthy of the subject.

We are grateful to many other people for their assistance in preparing this book. Antje Hammond has been indispensable in her contribution, in coordinating field work in Germany and Austria and assisting with translation of archival documents. We are also indebted to Heinz Manthey for his candour about the complexities of Elvira's story, and his own. Ute Hoffmann was most generous with her time in providing insights into her dealings with Elvira and the activities of Irmfried Eberl in Bernburg. Christoph Hanzig gave invaluable assistance with archival research at the *Sächsische Hauptstaatsarchiv* in Dresden, as did Peter Braun at the *Archiv des Bezirk Oberbayern* in Munich. Rolf Brüggerman, Peter Blum and Frank Pfenning at the MuSeele at the Christophsbad *Krankenhaus* in Goppingen provided us with access and guidance with both our field work and archival research, as did Günter Goller at the *Psychiatrie Museum kbo-Isar-Amper-Klinikum* in München-Ost. Friedrich Hauer assisted with our field work conducted at the site in Görden, as well as important archival research on the crimes of Julius Hallervorden and Hans Heinze. Ingo Wille from the *Stolpersteine* initiative in Hamburg generously shared the documents on Sonia Wechsler he has collected.

Acknowledgments

Many thanks also to our colleagues at various memorial sites for their insights and access to their artefacts and archival collections, including Hagen Markwardt at Pirna-Sonnenstein, Thomas Stöckle and Franka Rößner at Grafeneck, Peter Eigelsberger and Christa Memersheimer at Hartheim, Thomas Wieder at Hadamar, and also Herwig Czech at Am Spiegelgrund and Ulrike Hauffe at the *Justizpalast* in Nuremberg. Marek Rogowski assisted with our field work at the Treblinka memorial and sites around Lublin and Western Poland.

Frank Schneider was most generous with his time in relating his experiences and helping elucidate the German psychiatric profession's engagement with its past. Paul Weindling provided important insights into the complexities of the Max Plank Society's reconciling with its history in the National Socialist period and up to the present day. Miriam Wiersma provided helpful advice about the euthanasia statistics in Europe and United States. Linda Shields assisted with a perspective from her expertise in the history of the nursing profession in Nazi Germany at that time. Konrad Kwiet, distinguished Holocaust historian at The Sydney Jewish Museum, assisted with many introductions and expert advice on the project. We also thank Amanda Tink for reminding us of the work of Australian poet Les Murray and its significance to the topic. We are grateful to Mr Murray's Australian agent, Margaret Connolly, for the kind permission to reproduce his astonishing poem in the preface of the book, and in the book's title. We note with sadness Mr Murray's death just prior to this publication.

We are especially grateful to Scott Abbott at UTS ePRESS for his support, encouragement and invaluable advice that helped create this book.

MR and EL are grateful to their excellent colleagues at Sydney Health Ethics at the University of Sydney – Garry Walter, Wendy Lipworth and Miles Little.

The First Into The Dark

For inspiration, we are stylistically indebted to American author Adam Cohen, whose book *Imbeciles: The Supreme Court, American Eugenics, and the Sterilization of Carrie Buck* (New York: Penguin, 2016) traces the story of a young woman, Carrie Buck, to outline both the complexities of the American Eugenics movement of the early twentieth century and the operation of the US Supreme Court. His use of narrative biographies for the central people in that work enabled him to take his readers on a wider journey than might have been possible with a more detailed and dense history.

Ultimately, our aim has been to remember the victims of the *Krankenmorde* and commemorate their suffering. In order to contribute to this important process of memory and commemoration, we seek to assist the reader to both situate these crimes in the complex and turbulent history of mid-twentieth century Western culture, and to understand how important their legacy is for facing our current moral challenges.

Michael Robertson & Edwina Light – *Sydney*
Astrid Ley – *Berlin*

PART I
THE CRIME

Uchtspringe.

The name was branded irretrievably upon my soul. It wouldn't leave me in peace. It cast a murky shadow on every solitary thought and moment. It tormented me. The name meant only deep-seated fear; fear of a breed hitherto unknown; a black-masked dread that had barricaded me behind walls of icy stone. It had wounded me and turned me against mankind, against individuals who could turn a blind eye to the torture of innocent young children.

E Manthey (1995), *Die Hempelsche - Das Schicksal eines deutschen Kindes, das 1940 vor der Gaskammer umkehren durfte*
Lübeck: Hempel-Verlag Heinz Manthey.

PART I
INTRODUCTION

In his poem *Dog Fox Field*, Australian poet Les Murray powerfully alludes to the children assessed by doctors for 'feeblemindedness' as part of the Nazi eugenic 'cleansing' process. As the poem and the earlier Nuremburg trials revealed, it was children living with autism, Down syndrome, intellectual or other disabilities that were among the 'first' killed by the Nazi regime. In assisting their murders, the German medical and health professions handed the Nazi regime a model that could be applied for the 'darkness' of the ensuing Holocaust.

Darkness is a potent metaphor for the subject matter of this book. In particular, the darkness of the Nazi period in European history, with its millions of victims. Within this grotesque, uncomfortable darkness lies the *Krankenmorde*—the murder of the sick—wherein hundreds of thousands of people living with disabilities and illnesses were subjected to what would become known as the 'euthanasia program'. Traditionally interpreted as 'the good death', the use of 'euthanasia' as a euphemism for mass murder characterises the deceit necessary to perpetrate a crime of such scale. While occurring at the same time as the persecution of other groups in society, including lesbian, gay, bisexual and transgender people, Sinti and Roma people, Jehovah's Witness and other 'undesirable groups', the *Krankenmorde* would preface the monumental events of the Holocaust and the further murder of millions of people from Europe's many Jewish communities.

Among the first public policies of the Nazi government was the identification and forced sterilisation of a defined genetic underclass. This persecution soon transformed into a highly coordinated, medicalised and state-run program of mass killing. Those who lived with disability were the first into the darkness: the darkness of desolate care institutions and the gas chambers of the 'euthanasia' killing centres and the darkness of a 'modern' society and state that created such inhumane sites.

To explore the evolution of the *Krankenmorde*, we focus on the involvement of many individuals—perpetrators, victims, witnesses, those who aided the perpetrators, those who objected and opposed. Large scale historical moments, their complex origins, course and consequences demand deep and scholarly engagement, and this era infamously delivers precedents and dilemmas that continue to resonate. By considering the individual stories in their historical and social context we may better understand the ideologies, institutions and practices that lead to them—and perhaps better understand how they continue to flow today.

For their particular relevance we will follow the lives of the sisters Elvira and Lisa Hempel. In the way that many school children and adults learn about the Holocaust through direct testimony of victims, the histories of Elvira and Lisa Hempel allow us to bear witness to the mechanisms of the *Krankenmorde* and other forms of exclusion and deprivation at the edge of society during this period. Their personal journeys bring us into contact with many key people and places associated with the *Krankenmorde*. In following these testimonies we have in many cases adopted the terminology of the time, accepting that these reflect personal views and idiosyncratic mannerisms, including grammatical preferences which present some inconsistencies around spelling and correct translation from German to English. An early example is Elvira's childhood use of the term 'Totenmann' which would correctly be 'Totbringer', both referring to Death Man. In order to narrate the history of this time, many of the labels we must use in the text are jarring and perjorative and, where possible, we have utilised the contemporary terms that are used by different groups in referring to themselves.

Unique in many ways, universal in others, these accounts give us deeper insights into those historical events, as well as the individual experiences of some of the thousands of adults and children who were murdered because of judgments made about illness and disability.

The poem also alerts us to a darkness in our culture that might persist and return. In examining what happened in the *Krankenmorde*, we will also explore the repercussions and how these continue to evolve in contemporary scenarios of eugenics, euthanasia, and disability, and the responsibilities of groups and societies to remember and atone for the past.

CHAPTER 1
THE IRON DOOR

In the autumn of 1940, four ageing buses drove through the gates of the Uchtspringe State Hospital. Like other public institutions built at the end of the previous century, the hospital comprised numerous imposing buildings set within an extensive parcel of rural land. Originally positioned to take advantage of the estate's pleasant blend of meadows, farmland and forest, and with access to the river Uchte and a new railway station in nearby Uchtspringe village, the hospital founders had aspired to create a pioneering therapeutic project in psychiatry that was connected to local communities.[1] Uchtspringe was one of many small hamlets scattered through the wooded and gently rising hills, with farms and fields and stretches of pasture connecting the valleys and low lands beside the river and adjoining streams. Stendal, the main regional town, was some 25 kilometres to the north; Magdeburg, the closest city, was 60 kilometres south; and Berlin—one of the truly international and influential cities of Europe—was around 150 kilometres to the east. At the turn of the 20th century, secluded, peaceful, rural Uchtspringe seemed an ideal setting for the care and treatment of people living with physical, psychiatric and intellectual disabilities.

But much had changed since the rise of the National Socialist (Nazi) Government. By 1940 many of the Uchtspringe Hospital residents had been reassessed and sequestered, confined to certain building zones or locked inside the institution's now cold and squalid rooms for much of the time. Since the Nazis had taken power, those Germans too sick or disabled to contribute to society and the war effort were intentionally excluded and isolated from the rest of the community. The economic downturn of the 1930s had forced many unskilled people into working in institutions like Uchtspringe, and many lacked compassion for those who were hospitalised or interned there. The buses that motored through the grounds were almost identical to those used for postal services in Germany at the time; but with their windows painted over, they too had been modified for new purposes.

One of the four buses had been allocated for the residents of *Haus* 50, a wing of the hospital that had been home to hundreds of children with disabilities. This *Kinderfachabteilung* (children's ward) was now all but empty. Among the remaining few was eight-year-old Elvira Hempel, who had been declared 'feebleminded' and placed there in 1938. Local authorities had taken Elvira into custody when she was five, having decided that her family could not care for her. At Uchtspringe she joined her younger sister Lisa, who had been interned there soon after she was born in August 1935. While Lisa was a seemingly healthy newborn child, her mother had refused to take her home, telling hospital staff that she could not care for another child. Lisa was placed in institutional care as an infant, never to live with her family.

Within *Haus* 50 was a special ward for newborn babies and infants where, once fed and changed, they were often left alone in their cots, screaming behind locked doors.[2] On the day the bus came for Elvira there were no more babies in the special ward; bare mattresses were the only evidence that they were once there.³ The ones who cried too much, or made strange noises, had fallen silent after a nurse with a syringe went to their cots. Later, *Totenmann*, the man with the white sheets, would come and wrap the babies and take them away. The children of *Haus 50* feared *Totenmann*, the 'death man'.

Earlier that morning, a nurse had ordered Elvira and the other eight children to change into new clothing and be ready to leave. Elvira's much-loved floral dress was replaced with a different dress, a red one with many buttons. Elvira thought it ugly and did not want to wear it, but she knew she would be beaten if she did not obey.

Once dressed, the children were marched to another building nearer to the waiting buses and into a large hall where an elevated stage at the front held a long table overloaded with files. Piles of documents and discarded clothing were on the floor. There they joined a throng of elderly women and other children, gathered together in the hall like an audience, waiting.⁴

After a discussion among the nurses the doors of the hall were opened to the four buses waiting outside. The children were told to go first and were loaded on to a bus, one child to a seat. The bus had its windows blocked out from the inside with blue paint. Using her fingernail, Elvira secretly scratched a small hole in the paint to look out.

The bus pulled out of the grounds of the Uchtspringe hospital and drove along through a series of small towns. From her spy-hole Elvira could see trees lining the streets and then houses sitting in large fields in farmland. The trees still held their green summer leaves and here and there were glimpses of people at work in the fields or making their way through the streets of the towns and villages. The area seemed untouched by the war. They crossed a wide river and later entered a large town along a busy road. Soon after, the bus passed through a gate guarded by men in uniforms, perhaps soldiers or policemen, before slowing and stopping in a courtyard. Elvira could see nearby a large, brown brick building, three stories tall, with bars on the windows.[5] Behind this was a single story building that looked like a barn. On the opposite side of the courtyard was a church and other outbuildings.

The side door of the bus was pulled open and a woman climbed in calling, 'Raus! Raus! (Get out!)'. The children were hurried into the barn-like building, down a narrow corridor and into a series of dark connected rooms with no windows. Here, in the yellow glare of electric lights, Elvira saw piles of clothes and shoes, and beyond, a table laden with files. The adults in the room were dressed in white uniforms and looked like doctors and nurses. At any given time four of them would sit at another table set at an angle in the back corner of the room, consulting files. As she looked further, Elvira saw an iron door with two bolts set in one wall, like the kind used in air raid shelters. With the other children pushing into the room behind her, forcing her further inside, Elvira found she was standing on her own behind a large man near the piles of clothing and shoes. She decided it would be best to stand very still, to try and be invisible.[6]

Some of the women in the room were yelling at the children, 'Get undressed! And hurry up!'. After the children did so, a woman would grab each naked child by the back of the neck and stand them in front of the adults sitting at the table. Without explanation, one child after another was then taken into the room behind the iron door. A few of the children could not remove their clothing and the women became impatient, wrenching at the clothes and pulling them off. Elvira saw a woman grab the smallest child, a boy, and 'with one yank up and one down' he was naked.[7] She held him by his upper arm, first in front of the table and then to haul him the next few steps to the iron door. The boy was panicking and wanted to escape. Elvira saw him moving his legs to run, but they just thrashed in the air as he hung from the women's hand before he was hurled behind the door.

Soon all the other children had been processed. Elvira wondered if she had been overlooked, standing fully dressed behind the large man and staring at the growing piles of clothes and shoes. But suddenly the attention shifted to her and the woman marched toward her, 'Undress right now, or you'll get a beating'. Elvira did so carefully, her fingers unpicking each button of the dress. She was conscious of being watched and nervous of the punishment she would receive in this place. She took off her dress and threw it on the pile of clothes. She untied the laces of her shoes slowly, trying to delay whatever was coming next. But soon they were off and thrown on the pile of other little shoes. Now naked, Elvira felt a hand grab her left arm. She was dragged in front of the people at the table.

A severe looking man in a white coat demanded, 'Name?'

'Elvira Hempel.'

'How old are you?'

'Eight.'

He looked at a file and hesitated. He looked back at her, perplexed, then said: 'Get dressed again'.[8]

She did as she was told, finding her clothes still on the top of the pile. She dressed but could not remember anything more. All the other children remained locked behind the iron door; the adults in the room continued with their tasks.

Elvira was taken away to the tall, brown brick building she had seen on arrival and led upstairs to a prison cell. There, she and two other girls would be held for more than a week. The cell was locked each evening and unlocked in the morning, but the girls could not leave the prison ward. A woman brought them food every day yet no water was provided to wash or brush their teeth, and all they could do was play with the enamel chamber pots or explore the long corridor outside. As the days continued the children felt increasingly isolated and neglected; but at least no-one is beating us, Elvira thought. When she discovered a familiar looking scarf in an empty cell nearby, she was mystified. The blue-grey scarf looked like one worn by one of the elderly women back at Uchtspringe. Elvira knew this place was a prison. But why would a granny be imprisoned? Why would an eight-year-old girl be in prison? What had they done wrong? What were their crimes?

After breakfast on the eighth day, another woman came for them. She took the girls to a waiting taxi and they were driven 6 kilometres to a psychiatric hospital in the town of Görden. There Elvira would be held in the paediatric psychiatric ward until March 1941 when she was sent to an asylum in Altscherbitz (northwest of Leipzig). A few months later, in May 1941, she was returned to Uchtspringe, the place she had come to fear and loathe. Abandoned by her family and witness to the abuse and beatings in *Haus* 50 she yearned to be reunited with her sister, Lisa, who had been taken away with other children in the weeks before Elvira left. As she tried to understand her predicament and find ways to navigate through the physical and mental distresses of the psychiatric wards, she was sustained by the hope that Lisa might also be returned to Uchtspringe.

Thus was the child Elvira Hempel pushed along the margins of the Nazi's medical 'euthanasia' system. Deeply traumatised, but by no means 'feebleminded', Elvira would experience far more in the years ahead. Yet she would often be drawn back to the concrete room, the piles of clothing and the nurses pulling the handles shut on the iron door. What had been her crime? Why were the others sent behind the iron door, yet she was not?.[9]

FIGURE 1 The Uchtspringe Institution in 1903

FIGURE 2 The Uchtspringe Institution in the 1940s

Elvira Hempel was born in October 1931 in Magdeburg, and named after her mother, Elvira-Lotte. Elvira-Lotte would give birth to 15 children, although only six would survive infancy. Even after leaving her infant daughter Lisa in the care of authorities, so overwhelmed was Elvira's mother with caring for her brood in those difficult days of the Great Depression, that she would tie the legs of the younger children to the dining table while she went out to find work or food.

Elvira's father Otto was a habitual and recidivist criminal. His offences included rape, burglary and fraud.[10] Despite his unruly behaviour, Otto was something of a local character in Magdeburg, busking with his accordion in the town's streets with one of his children singing along. Elvira remembered that he was otherwise ashamed to be seen in the company of his family. He was, however, happy to involve one of her older brothers, Otto junior, in his schemes.

Elvira learned from her mother that Otto had squandered a small inheritance and often wasted any money he had procured on petty indulgences, particularly in seeking the company of women. The family defaulted on rent payments on a few occasions and were put out on the street. Under such financial duress, Elvira and her siblings regularly picked through rubbish at a nearby dump to find scrap metal to sell. One of Elvira's clearest memories of childhood was her father catching an Alsatian dog and slaughtering it in front of the family to provide food for a few days.[11]

The Hempel children were often beneficiaries of charity from either Protestant or Catholic churches; they learnt quickly to present themselves as new converts to different congregations in the town to obtain food and clothing.[12] The family were generally considered pathetic figures within the community; when the children did attend school, the better-off families held food collections to try and help them. Other residents of the town were not as considerate, labelling the family, particularly Otto, as 'undesirables'. Their house was fire bombed on one occasion after Elvira-Lotte went to social services seeking help.[13]

When Elvira's five-month-old brother Heinz died in October 1934, the family could not afford a funeral. Elvira recalled that Otto stole an infant's coffin from a nearby church. Elvira watched as Otto forced the boy's body into the tiny casket. He then took it to the nearest church cemetery and abandoned it there in the hope it would be buried.

Throughout childhood Elvira was plagued by what was thought to be eczema, although when treated she was considered contagious, so it is possible these recurring skin infections were chronic impetigo (streptococcal skin infection). On many occasions she was hospitalised with diffuse eruptions of severe dermatitis. Because of these frequent hospitalisations and periods of quarantine, her formal education was limited to barely two years of school. By 1936, social services became sufficiently concerned about Elvira, now five years old, to take her from hospital and place her into the care of a Catholic – run institution in Magdeburg.

In September 1938 the local authorities took Elvira to the paediatric clinic at Magdeburg-Sudenburg, where a child psychiatrist, Professor Dr Ernst Walter Fünfgeld, assessed her. The record of Fünfgeld's examination[14] noted Elvira was incontinent of urine at night, had little or no capacity for basic mathematics and 'plays with bricks like a toddler'. He noted she was 'aggressive towards other children in the institution' and that she 'lies and tortures other children without reason'. Fünfgeld proclaimed her *'unterwertig'* (mentally inferior) or what is now termed 'intellectually disabled'. Following Fünfgeld's assessment, Elvira was sent to the newly established *Kinderfachabteilung* (special children's ward) at *Haus* 50 in the Uchtspringe hospital.

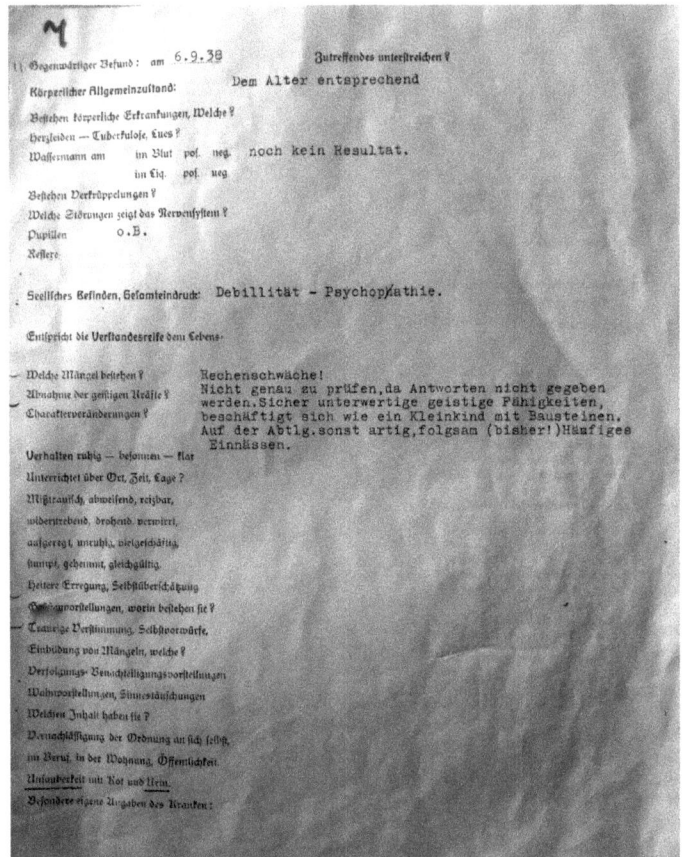

FIGURE 3 Notes from the assessment of Elvira Hempel by Professor Fünfgeld in 1938
It reads 'debility from psychopathy' and notes her problems with mathematics, 'playing with building blocks like a toddler' and writes 'surely is intellectually disabled'.

Elvira's first impression on arriving at Uchtspringe was that most of the adults and children seemed to be severely disabled. She recognised that she was being held with people who had many forms of physical and mental disability, including profoundly emotionally disturbed patients. Observing their treatment and care would shock her. Yet among these traumatic scenes, Elvira's moment of greatest joy occurred on her second day in *Haus* 50 when she was reunited with her sister Lisa, then aged nearly three. She had last seen Lisa as a newborn baby at Magdeburg hospital. Despite all that would happen to them, caring for and protecting Lisa became Elvira's mission.

As time passed, Elvira was determined to be active and avoid punishment. She sought to occupy herself and impress the nurses. She started by polishing the floor of the long corridor in the ward, up and down, over and over, anxious to please them so that they would allow her to continue this work, yet all the while terrified that she would be seized and beaten again.[15] The nurses let Elvira keep this daily task of polishing the floor, allowing her the freedom to roam and observe more closely what was going on at Uchtspringe. She would watch a nurse walk up the corridor with syringe in hand to the babies' dormitory, sometimes following her and peering in to see her injecting the babies. She would later pass the *Totenmann* in the corridor, taking away the wrapped bundles of dead babies on his trolley. Elvira also observed many 'old ladies' arriving at Uchtspringe, but before long they too were gone. Around this time many children, particularly those considered disobedient, also began to disappear from *Haus* 50.[16] One of Elvira's duties was to collect and sort the piles of clothes that were left behind. Later that year she would be confronted with the loss of her own sister when, just before her fifth birthday, Lisa Hempel was also taken away. Within a few months, Elvira and the last of her young companions would be ordered on to the bus.

Elvira's later accounts of her time in institutional care are dominated by the theme of bedwetting. In the first institution in Magdeburg, bedwetting children were isolated from the others;

beaten, ridiculed, and made every morning to wash and dry their sheets. Lisa Hempel, having never been nurtured or attached to a loving caregiver, was developmentally delayed and constantly wet her bed. She was routinely punished and mocked by the nurses because of her enuresis (involuntary bed wetting). To protect Lisa, Elvira developed a strategy of taking her sister into her bed each night, leaving the sheets in the other bed dry. The following morning she changed the wet sheets in her bed with clean ones from a closet near the dormitory.

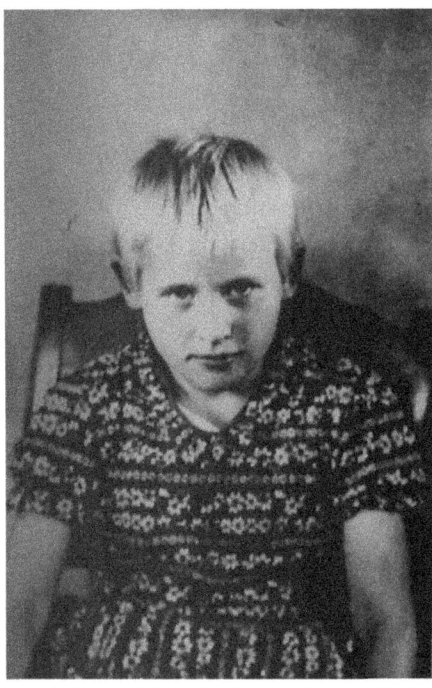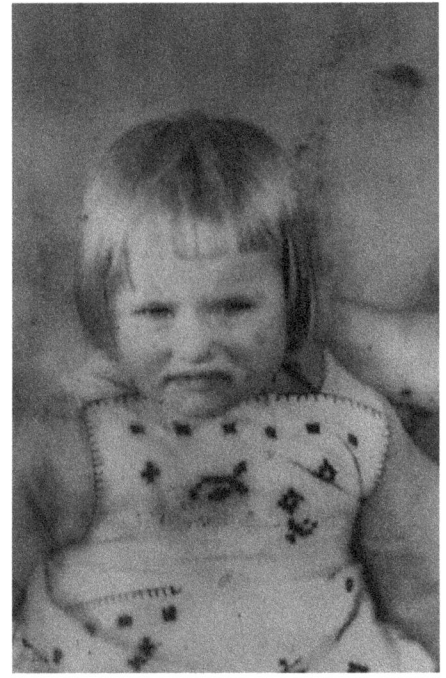

FIGURE 4   Elvira Hempel (left) and Lisa Hempel (right)
This photo was likely taken at the Uchtspringe Institution in 1940. Note the arm supporting Lisa to sit. Her gross motor development was so delayed she was unable to support herself.

We now know that Elvira and the children were transported from Uchtspringe on 3 September 1940 and taken to a 'euthanasia' centre in the town of Brandenburg an der Havel[17], one of six *Tötungsanstalten* (killing institutions)[18] that operated in Germany and Austria in the years 1940 to 1941. While most of these *Tötungsanstalten* were located in the buildings of former psychiatric hospitals or care institutions, the complex on Neunendorfer Straße in Brandenburg an der Havel had been originally a prison dating from the eighteenth century. As early as 1933, the new Nazi government used it briefly as a concentration camp.[19] Its high walls made it a suitable site for clandestine operations.

In late 1939 the SS Office of Budget and Construction took possession of the buildings for the purposes of establishing a *Tötungsanstalt*. Like the other killing centres in the 'euthanasia' program, the one in Brandenburg an der Havel[20] was equipped with a carbon monoxide gas chamber, first disguised as an inhalation room, where victims were told they would be breathing in concentrated oxygen to improve their health. Fake shower heads were later fitted to the ceilings, probably in March 1940. Measuring three by five metres, with a three-metre-high ceiling, the chamber was tiled and lined with benches running around the room's four walls. A pipe with holes in it, out of which the poison gas carbon monoxide streamed, ran along the wall at a height of about 10 centimetres. The carbon monoxide pressure cylinders stood outside the room. Set into the iron door that led to the gas chamber was a rectangular peephole through which the killing process could be observed. When the gassing was finished, the gas chamber was aired by means of a ventilation system.

The crematoria ovens, in which the corpses of the victims were incinerated, initially stood in a space directly behind the gas chamber in the prison barn, separated from it by solid double doors. To improve secrecy, the crematoria ovens were later moved to a remote plot of land in the village of Paterdamm, outside the town of Brandenburg an der Havel in the summer of 1940.[21] The new crematorium was disguised as the 'Chemical and Technical Research Institute'. The bodies of the victims were taken there at night by a vehicle disguised as a post office van.[22]

More than 9,000 people were murdered in Brandenburg killing centre's gas chamber.[23] The youngest victim was aged two, the oldest 87.[24] Brandenburg killing centre operated for just ten months—from January until late October 1940. The staff were then transferred as a group to a new killing centre in a psychiatric hospital located near the town of Bernburg, around 150 kilometres west of Berlin on the river Saale.

FIGURE 5 The Brandenburg extermination centre in the 1940s

The Nazi regime's policies that directed the Hempel sisters to the gas chamber at Brandenburg had complex origins in racial and eugenic theories, in the problematic social status of the psychiatric profession in Germany at that time, and in a phenomenon we now recognise as 'biopower'—a form of power applied by government to control the biological development of its population. These factors enabled the Nazi regime's program of persecution and murder of people living with psychiatric, intellectual and physical disabilities. By the end of the war around 360,000 German citizens deemed 'genetically unfit' had been forcibly sterilised, and approximately 300,000 people with various forms of illness or disability had been murdered in Germany and its occupied territories.[25] The mass murder of the sick and disabled proceeded under the euphemism 'euthanasia', although in present day Germany the term *Krankenmorde* (the murder of the sick) is now used to describe the crime.

The *Krankenmorde* involved the development and application of a coordinated, bureaucratic process requiring identification of victims, central registration, transportation to purpose built killing centres via intermediate institutions, and an elaborate deception of the victims and their families. In effect, this operation provided the model for the Nazi's 'Final Solution'—the planned mass extermination of Europe's Jewish population and many other 'undesirables'.[26] In this first phase of the *Krankenmorde* more than 2,000 victims were Jewish psychiatric or medical patients, making them among the first victims of the Holocaust.[27]

While Brandenburg and the other five killing centres had ceased operations by August 1941, the killing program was continued surreptitiously in hospitals in Germany, Western Poland and Austria. In these new killing centres, patients were murdered by various means, including deliberate overdose of medication or being starved to death in dedicated *Hungerhäuser* (starvation houses) through specific feeding regimes which gradually reduced caloric intake.

On arriving at the Hadamar 'euthanasia' centre in April 1945, Ray Leopold, a medic serving with the 112th Infantry Regiment of the US Army, reported that seeing the diseased and emaciated bodies of surviving asylum patients and the open mass grave nearby was by far the worst thing he had witnessed in the war.[28] Yet, for many, the extent of this systemised incarceration and murder would remain unknown until long after the end of World War Two. Even with the evidence arising from the Nuremburg trials, it would not be until the trial of Adolf Eichmann in 1961, and the broadcast of the television miniseries 'Holocaust' in the 1970s, that broad international attention would be drawn to the mass killings perpetrated by the Nazi regime. It remains difficult to comprehend the deliberation and resolve with which it was put into action. There were rigorous steps taken and willing collusion by the legal and medical systems that were authorised to enact the *Krankenmorde*, supported by the apparatus of the regime.

While German authorities sought to keep the *Krankenmorde* hidden from the German people, there was little doubt where the order had originated.

CHAPTER 2
'LIFE UNWORTHY OF LIVING'

In October 1939 Adolf Hitler wrote a missive to his personal physician Dr Karl Brandt. It was back-dated to 1 September 1939, the day Germany had started the war:

> '*Reichsleiter Bouhler and Dr. Brandt, M.D. are charged with the responsibility of enlarging the authority of certain physicians to be designated by name in such a manner that persons who, according to human judgment, are incurable can, upon a most careful diagnosis of their condition of sickness, be accorded a mercy death.*' [29]

With this brief written order, the Nazi state was set in the direction of the *Krankenmorde*. The Nazi policy that contrived to put Elvira and Lisa Hempel to death at the Brandenburg killing centre never existed in law. Hitler's order—one of the few recorded in writing during the Third Reich—allowed the establishment of a determined bureaucratic system that would coordinate the functions of the state to ensure that all 'undesirables' were identified, assessed, removed and, where necessary, 'exterminated', as efficiently as possible.

Yet the preconditions to this state-run 'euthanasia' program did not exist solely in Nazi Germany. The popular appeal of eugenic theory had spread across many countries and preceded Adolf Hitler's program of mass-murder of the sick and disabled by many years. Indeed, given the enthusiasm for eugenics, compulsory sterilisation of the 'genetically inferior', and the alienation of people living with mental illness and disabilities in the United States and numerous countries across Europe and the Anglosphere during the same period, one is left to consider why such crimes were confined to the Nazi state alone.

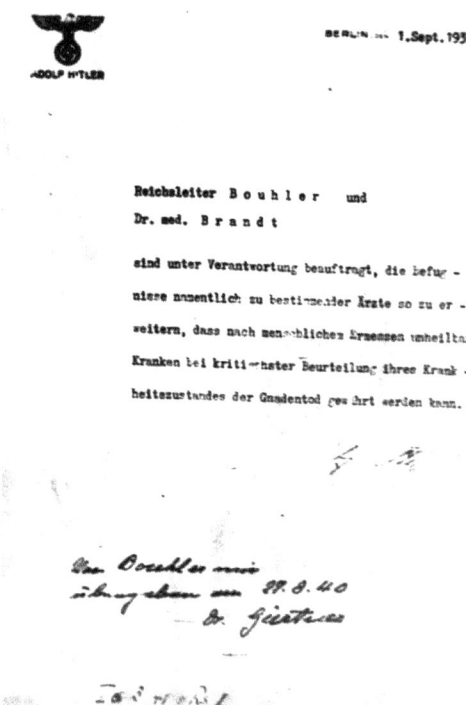

FIGURE 6 Direction from Adolf Hitler to Philipp Bouhler and Karl Brandt, 1 September 1939
The letter from Hitler to Bouhler and Brandt backdated to 1 September 1939 authorised 'certain physicians' to accord a 'mercy death' to 'persons who, according to human judgment, are incurable'.

By the end of the 1914-1918 war, Germany had lost more than two million men.[30] In the opinion of many politicians and physicians, the best of German manhood, and by extension the best genetic stock, lay dead in the battlefields of France and Belgium, while 'inferiors' remained and procreated in Germany, condemning her to a bleak future. The humiliation of defeat and the crippling economic penalties imposed on Germany in the Versailles treaty prompted economist John Maynard Keynes in 1920 to warn of future geopolitical disaster for Germany and Europe.[31] Keynes' sagacity about a future revanchist Germany contrasted with his eugenic views, which included arguments that states should regulate both the size and 'quality' of their populations.[32] He was part of an extensive and influential international movement comprising medical and health practitioners, academics, artists, scientists, lawyers, politicians, industrialists and wealthy patrons with similar eugenic views and he would later become Director of the British Eugenics Society (1937-1944). Almost as predicted, economic and political instability doomed the fledgling Weimar republic and enabled populism and political extremism to gain traction within Germany. According to a myth palatable to the extreme political right, Germany lost the war because she was 'stabbed in the back' by Jews and Bolsheviks.[33] When appropriated by the Nazis, the racial themes of this myth became the organising principle of their project for Germany's future as a utopian racial state, cleansed of racial and hereditary 'inferiors'.

In 1920, Karl Binding, a highly respected retired jurist, and a psychiatrist, Alfred Hoche, published a monograph titled *Die Freigabe der Vernichtung lebensunwerten Lebens* ('Allowing the Destruction of Life Unworthy of Living').[34] Binding had served as an orderly in a field hospital during the Franco-Prussian war in 1871 and later lived where he worked in the law in Leipzig, Germany. Hoche had worked as a researcher in neurology and had lost a son during the 1914-1918 war in Belgium. Binding and Hoche's monograph posed the question of whether killing a terminally ill patient, at his or her own request, should exempt someone from criminal punishment. Although they emphasised that *Sterbehilfe* (assisted dying) should never be carried out against a competent patient's will[35], Binding and Hoche expanded their concept of euthanasia into three defining groups. First, those severely injured or terminally ill, who both understood their situation and wished to end their suffering but could not express such a wish. Second, those who were competent but comatose and would suffer 'nameless misery' if woken.[36] Third, those 'incurable idiots who are the terrible image of real people and cause horror in each person who faces them'.[37]

Binding and Hoche argued that the value of an ill or disabled individual was diminished if his or her social contribution to the nation was outweighed by the expense of caring for them. In their terminology, such people were 'useless eaters' whose life was a 'ballast existence'. The idea of killing people deemed useless for the society was, however, not new. The German biologist and philosopher Ernst Häckel had first expressed it with respect to the 'genetically inferior' in his book *History of Creation* in 1868.[38] Häckel's biogenetic theory had a profound influence on Western culture[39], including the evolution of Nazi racial and eugenic philosophy. As evolutionary scientist Stephen Jay Gould would later observe, Häckel's ideas 'buttressed racism and its colonialist extensions'[40] with its hierarchy of lower and higher races.

These ideas were to coalesce with the philosophies of eugenics and social Darwinism. The term 'eugenics' usually refers to the improvement of the 'genetic health' of a population by the promotion of desirable heritable characteristics. This is realised through the alteration or elimination, by whatever means, of 'inferior' genetic stock from the breeding pool.

Although the concept dates to Plato[41], the term 'eugenics' is usually associated with the ideas of British polymath Francis Galton.[42]

Through the second half of the nineteenth century, Galton sought to build on the observations of his more famous cousin Charles Darwin and explored theories about improving the human species through patterns of selective breeding by encouraging appropriate marriages between humans deemed to have desirable genetic features. In the first few decades of the twentieth century, groups interested in 'eugenic science' flowered in numerous countries and grew in popularity, in part due to significant societal change and political unrest that followed the 1914-1918 war, and the economic turmoil of the Great Depression. Eugenic ideas were used to legitimate racist immigration and other social policies in the 1920s and 1930s. Eugenic societies became influential in numerous countries, attracting high minded and usually wealthy white Christians with concerns about immigration, population density and mix, and 'aberrant' behaviours such as alcoholism, prostitution and 'lunacy'. This developed into a determined movement for social change in the United States under the influence of the prominent American eugenicists Charles Davenport, Madison Grant, Lothrop Stoddart and Harry Laughlin. American eugenic institutions received endowments from wealthy benefactors such as the Carnegie, Rockefeller, Kellog and Harriman families. The prolific journalist and British physician, Caleb Saleeby, professed views on race and eugenics that were influential on both sides of the Atlantic. To popularise eugenics outside of academia and social elite circles 'Better baby contests' were held at many state fairs to educate the public about raising healthier offspring.[43]

The US state of Indiana passed the world's first compulsory sterilisation law in 1907 and, until the 1960s, there were 29 other US states with similar legislation. Eugenic sterilisations were enforced in the US until the early 1960s by which time more than 60,000 Americans had been forcibly sterilised.[44] Most victims were women and enforced sterilisation of Native Americans continued until the late 1970s.[45] These sterilisation laws were the focus of the infamous 1927 US Supreme Court case *Buck-v-Bell*, where, in assenting to the forced sterilisation of patients in mental institutions, Chief Justice Oliver Wendell Holmes Junior proclaimed that 'three generations of imbecile is enough'.[46] The 'euthanasia' of those with severe disabilities was also debated in the United States at the same time as in Germany. Cornell University professor of neurology, Robert Foster-Kennedy, argued in 1942 that feebleminded children should be euthanised as a matter of public policy.[47] Foster-Kennedy's view was provocative and enjoyed some support among his colleagues in the American medical profession, although when the process of the killing of the disabled under Nazism became apparent, German and American eugenic movements parted company.[48] It has been claimed that the primary reason for this historical divergence was the moderating influence of the US federal government and the plurality of views within American eugenics.[49]

In Germany, many in the community and the professions endorsed Binding and Hoche's arguments. In 1920, soon after their book had been published, Ewald Meltzer, the director of an asylum for 'feebleminded' children in Saxony, conducted a survey of 200 parents of the children under his care[50] asking if they would agree to a 'painless curtailment of [the] life' of their child, if such a law existed. Of the parents who responded, more than 70 per cent approved the idea provided that experts had established their child was suffering from 'incurable idiocy'. Among the parents who opposed the proposition some indicated they would have agreed to the child's death in circumstances where the child became an orphan.

After the Nazi regime took power in Germany in January 1933, newly issued school textbooks posed mathematics questions to students, asking them to calculate the cost to the state of keeping a mentally ill person alive in an asylum[51] or of housing genetically defective families.[52] This kind of utilitarian abacus figured in the later justifications for the mass murder of those who were considered 'ballast existence'. The 'removal' of the 70,000 'useless eaters' in the official 'euthanasia' killing program was later calculated to have saved Germany 245,955,000 Reichsmarks (RM)[53] per day over 10 years, including RM 14,420,023 in potatoes, RM 1,054,080 in cheese and RM13,281,606 in vegetables.[54]

In July 1933 the Reichstag passed the *Gesetz zur Verhütung erbkranken Nachwuchses* (Law for the Prevention of Genetically Diseased Offspring) legislation inspired by the compulsory sterilization laws in the United States.[55] The German law listed eight illnesses regarded as hereditary: five psychiatric diagnoses, including intellectual disability, schizophrenia, epilepsy and manic-depressive disorder; two physical disorders; and severe alcoholism. Huntington's disease, listed specifically in the law, was the only demonstrated hereditary disorder; all the other legislated disorders had a speculative genetic basis.

Unlike the later 'euthanasia' program, this compulsory sterilisation was performed on the grounds of a State law and consequently there were no attempts by the Nazi regime to hide what was happening. Men usually underwent vasectomy, while women were subject to tubal ligation. Other less common methods of attempted sterilisation included x-ray and radium irradiation. To legitimate the process to a wary public, decisions for or against state enforced sterilisation were transferred to newly established 'Hereditary Health Courts', ostensibly set up to examine every case. Almost 200 Hereditary Health Courts were eventually established in the German Reich. Each comprised three members: two doctors and a judge. A second appellant court heard appeals against a judgment of the lower court.⁵⁶ Around 90 per cent of cases referred to the appellant courts proceeded to sterilisation.

Despite many legal appeals, by the end of the war in 1945 nearly 360,000 people had been sterilised against their will. More than 5,000 people died because of sterilisation operations.⁵⁷ The high rate of 'sterilisation sentences' ⁵⁸ was the result of special provisions for these court proceedings which left those who had been designated with a 'disorder' little chance of avoiding sterilisation through legal argument or appeal.⁵⁹

Outside this formal court system, these laws were used from 1940 to justify sterilisation of Polish and Russian *Ostarbeiter* (forced labourers) who were diagnosed with hereditary disorders or mental illnesses.⁶⁰ Sterilisation laws were also used against other people on racial grounds, such as Jewish, Sinti and Roma people.⁶¹ A little acknowledged instance of the illegal application of Nazi sterilisation policies involved the *Mischlinge* (mixed-race children) of German women and French colonial troops from West Africa, garrisoned in the Rhineland after the German defeat in 1918.⁶² These children, the so-called 'Rhineland Bastards', were an affront to the Nazi regime primarily on grounds of racial, as against genetic, hygiene. The Nazi government reoccupied the Rhineland in 1936, after which it established a Commission under the direction of Eugen Fischer of the Kaiser Wilhelm Institute of Anthropology, Human Heredity and Eugenics. In 1937, around 400 of these children were sterilised in a secret campaign.⁶³

―――

By 1940, under Hitler's 'mercy death' decree, a physician's medical judgment now put many thousands of patients at risk. By then imprisoned at Uchtspringe, Elvira Hempel's fate was sealed by the diagnosis of 'feeblemindedness' made by the Magdeburg child psychiatrist Dr Fünfgeld. This term was not confined to the eugenic discourses of the early twentieth century. In the Christian New Testament (King James version, 1611), the First Book of *Thessalonians* (5:14) urges readers to 'Comforte the feble mynded'. Two millennia later the writer Jack London's confronting short story 'Told in the drooling ward' (1914) makes light of the behaviour of people with intellectual disabilities in institutional care in a California asylum. In the eyes of his protagonist, London seems to hold a mirror to society's disparaging attitude towards the 'feebs'.[64] The clinical construct of feeblemindedness seems to date from the mid to late nineteenth century. In August 1887, a presentation at the National Conference of Charities and Corrections in Omaha, Nebraska, outlined a definition of 'feeblemindedness' as an umbrella concept incorporating a heterogeneous group of people with varying levels of intellectual disability, much like the modern term 'developmentally delayed' or the now jarring 'mentally retarded'.[65] Feeblemindedness became a primary preoccupation of eugenicists and would soon be linked to other socially 'undesirable' traits such as criminality, sexual promiscuity and 'work shyness'.

The term 'feebleminded' is often associated with the American eugenicist and psychologist Henry Goddard. In 1912 Goddard published a persuasive, but now discredited, study of the illicit offspring of a hero of the American Revolutionary War, Martin Kallikak.[66] Based upon the presumed genealogy of Deborah Kallikak, a 'feebleminded' woman in the Vineland Training School for Feebleminded Children, a medical institution in New Jersey, Goddard surmised that her great-great-great grandfather Martin Kallikak had sired the ancestors of the 'feebleminded' family proband Deborah,[67] after a dalliance with an unknown barmaid on his way home from the war. Martin Kallikak later married a Quaker woman with whom he produced ostensibly 'normal' offspring. Goddard applied very basic laws of Mendelian inheritance[68] to the intellectually disabled limb of the Kallikak family tree and concluded that this provided strong evidence for the heritability of 'feeblemindedness'. Goddard's book was profoundly influential in the United States, effectively inspiring a raft of involuntary sterilisation laws in different states. In Germany, his Kallikak family tree was included in a series of displays on the topic of inheritance and racial hygiene presented by the prominent Dresden German Hygiene Museum in 1923.[69]

FIGURE 7 Deborah Kallikak at work in the sewing room of the Vineland Training School for Feebleminded Children, c1910
An image from Henry Goddard's study *The Kallikak family: A study in the heredity of feeble-mindedness.*

Goddard also appropriated the work of French psychologists Alfred Binet, Victor Henri and Theodore Simon, who in 1905 had developed a means of quantifying intelligence by focusing on verbal abilities based on chronological age. This research had sought to establish reliable means of identifying children with learning disabilities in French schools. Their hope was to help refine teaching methods to assist these children in class, rather than the usual approach of excluding them from education altogether. However, Binet was cautious about applying the test, stating that 'intelligence' was a complex phenomenon, better studied qualitatively.[70] Goddard adapted Binet's work for his agenda in 1908, later publishing his own version of the psychometric instrument measuring 'Intelligence Quotient' (IQ). His work resulted in a program of testing for IQ being introduced into American public schools in 1911. Goddard proposed an IQ-based taxonomy of intellectual disability: 'morons' (IQ of 51-70), 'imbeciles' (26-50), and 'idiots' (0-25). Goddard considered morons or those of lesser intellectual capability as socially unfit, necessitating exclusion from society, compulsory sterilisation, or both.[71] In Germany, the term *Idioten* had come into use to describe the same population of people with intellectual or learning disabilities. By 1913 Goddard had convinced US authorities to utilise his IQ testing on prospective immigrants held on Ellis Island. The tests were not modified for linguistic and cultural difference, resulting in a majority of those tested being assessed as having low IQs. This unchallenged fallacy would become supporting evidence for the eugenicist's sterilisation campaign.

The international eugenics movement held three international conferences between 1912 and the Great Depression, and saw flourishing international collaboration between academic institutions in Europe, North and South America and the Antipodes. Apart from the US and Germany, eugenic sterilisations were enforced in the 1920s and 30s in Sweden, Japan, Canada, Brazil and many other countries.[72] Australian eugenicists focused their efforts at ethnic cleansing of the Aboriginal population through removal of mixed race children from their communities[73]; their attempt to persuade their federal government to introduce a compulsory sterilisation law was thwarted by the Great Depression.[74]

Through Goddard's efforts in the US, eugenics and racial hygiene were married together long before Germany's hereditary health law.[75] The catalyst for the Nazi regime progressing its hereditary health laws to the mass murder of the sick and disabled, was the murder of an infant boy in Leipzig in 1939.

―――

Richard Kretschmar was a farm labourer who lived with his wife, Lina, in Pomßen, a village south of Leipzig in eastern Germany. During the 1930s Richard and Lina became enthusiastic supporters of Nazism.[76] In February 1939 their son Gerhard was born blind, with phocomelia (malformed or absent limbs) and suffering seizures. His doctors suspected that he was also 'feebleminded'. Appalled by the condition of his profoundly disabled son, Richard Kretschmar soon referred to him as 'this monster'. When Gerhard was six weeks old, his father took him to the Leipzig Children's Clinic and asked the director, Werner Catell, to euthanise the baby. Catell refused as at the time such an action was illegal. Not satisfied, Richard Kretschmar next wrote directly to his country's *Führer* for support and guidance. His petition was referred to Department IIb (Petitions section) of Hitler's Chancellery (KdF). Adolf Hitler took a personal interest in the case and delegated his escort physician, Dr Karl Brandt, to assess the situation.[77]

Brandt, a tall, handsome and charming man, always impeccably dressed, had insinuated himself into Hitler's confidence, becoming his physician in 1934 and thereafter a close adviser. After examining Gerhard in the Leipzig clinic, Brandt contacted Hitler to confirm Richard Kretschmar's concerns. Hitler then authorised the clinic staff to euthanise the baby. On Tuesday 25 July 1939, Gerhard

Kretschmar was administered a lethal dose of the barbiturate Luminal (phenobarbitone).[78]

The 'mercy death' of the Kretschmar baby would be a threshold moment in the Nazi regime's attempted extermination of the disabled. Gerhard Kretschmer became the first victim of a program of large-scale paedocide perpetrated by the Nazi regime that came to be known as *Kindereuthanasie* (children's euthanasia). Through the *Kindereuthanasie* the aims of the *Krankenmorde* were set in motion by Adolf Hitler's order to Bouhler and Brandt four months later, gathering up and condemning the 'feebleminded', mentally ill, alcoholic, 'socially undesirable' and chronically ill. Amongst them were Elvira and Lisa Hempel.

CHAPTER 3
AKTION T4 – IMPLEMENTATION AND RESISTANCE

The imposing classicist style villa on Berlin's *Tiergartenstraße* 4 was owned by a Jewish family, the Liebermanns. Hans Liebermann senior, a successful textile manufacturer, had bequeathed the residence to his son, Hans junior, in the 1920s. Hans junior committed suicide after the anti-Jewish violence of 9 and 10 November 1938, known as the 'November Pogroms' or by the infamous Nazi term *Kristallnacht*. After this the villa was confiscated by the SS Main Economic and Administrative Office and in the summer of 1939 became the site of a bureaucratic process that would facilitate the murder of more than 200,000 people.

Following Adolf Hitler's authorisation to Karl Brandt and Philipp Bouhler, head of the KdF, to commence the 'euthanasia' program in October 1939, the Ministry of the Interior sent registration forms to hospitals and nursing homes across the German Reich. An accompanying leaflet specified which types of patients were to be reported. Specific criteria required reporting on certain medical conditions, the duration of the patient's stay in the institution, criminal history, employment history, and current work capacity. Jewish patients or those of foreign citizenship were to be registered, regardless of their clinical status. The completed registration forms were returned to the administrative headquarters for the 'euthanasia' program which had been established at the stolen Liebermann home at *Tiergartenstraße* 4, leading to the now notorious codename *Aktion* T4 (Operation T4).[79]

The planning and implementation of the *Krankenmorde* was carried out by Hitler's Chancellory (*Kanzlei des Führers* or 'KdF') in close cooperation with the Ministry of the Interior. In the summer of 1939 the KdF began the systematic registration of disabled newborn children and infants as a precursor to the program of *Kindereuthanasie* ('children's euthanasia'). At the same time, KdF functionaries prepared for an organised campaign of murder of certain groups of adult patients in German hospitals and care institutions. The killing operations were to be performed in specially equipped *Tötungsanstalten* (killing centres) using carbon monoxide gas as the killing agent.

The decision of whether the identified patient lived or was to be killed fell to a committee of medical assessors (*Gutachter*) at the Medical Office of *Tiergartenstraße* 4. Around 40 of Germany and Austria's most respected physicians and psychiatrists had been recruited for the task of assessing and selecting victims. So large was the number of assessments that the committee was divided into a committee of junior assessors (*Untergutachter*) whose work was overseen by more senior and experienced assessors (*Obergutachter*). The T4 medical committee was hidden behind a fake organisation, the Reich Cooperative for State Hospitals and Nursing Homes (*Reichsarbeitgemeinschaft Heil-und-Pflegeanstalten* or 'RAG'). All correspondence or documents related to the T4 medical committee was sent to the RAG.[80]

Each life or death decision was based on the defined but limited information provided on each individual registration form (known as '*Meldebogen 1*'), completed by the patient's doctor and sent to the T4 committee, in combination with a review of the patient's medical file. No victim of the T4 program was examined in person by the committee as part of the selection process. On the T4 forms, a panel of three experts would insert a symbol determining: '+' (death), '-' (survival), or '?' meaning the decision was referred back to the patient's doctor. For a patient to be selected for death, the form required three '+' marks. The question mark that appeared on Elvira Hempel's form—which spared her from the gas chamber at Brandenburg killing centre—would likely have come from this committee. The primary criteria for the T4 selectors in determining life from death was a person's work capacity, although other considerations such as 'curability' or whether a person had regular visitors may have been a factor in their deliberations.

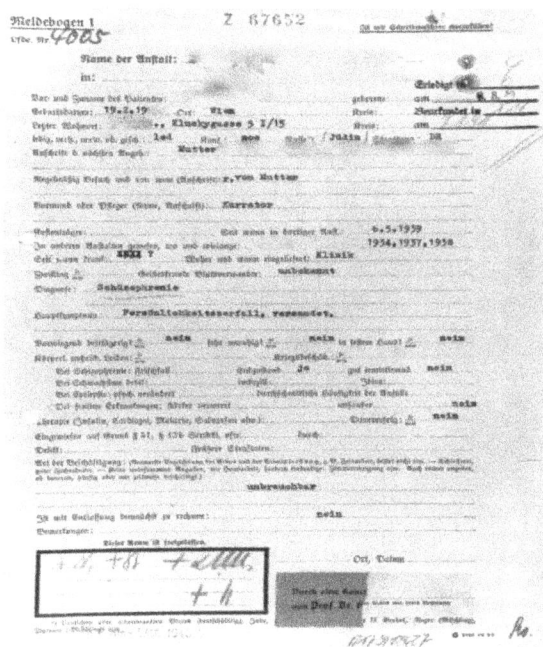

FIGURE 8
A registration card for the *Aktion* T4 program
These forms were completed by the victim's treating doctor and sent to the T4 medical committee in Berlin. The '+' marks in the bottom left corner indicate that the T4 assessors concluded the person would be given a 'mercy death'.

Patients condemned to death were then recorded in transport lists which ensured they were located and escorted to the killing centres. From spring 1940 onwards the victims were sent via various intermediate institutions called *Zwischenanstalten* (Intermediary institutions) enabling the regime to better organise and conceal their activities. A fake organisation *Gemeinnützige Krankentransport GmbH* (Charitable Ambulance) or *Gekrat* was established within the T4-apparatus, tasked with transporting victims to the killing centres. *Gekrat* made use of buses to transport victims, usually vehicles retired from the postal service that were then painted grey.[81]

On arrival at the killing centres, the victims were directed to undress and were then photographed and examined by a 'euthanasia physician'. The victims were usually told they were to take a bath or a shower. Once they were locked inside the gas chamber, carbon monoxide would be released from external cylinders into a pipe system inside the vault. As the gas was odourless, the victims usually had no sense of what was about to happen. After a short period some lost consciousness and then succumbed. Others panicked, screamed or rushed for the door as they struggled to breathe or realised they were dying. The euthanasia physician could observe through a peep-hole in the gas chamber door whether all the victims had died and, if so, the gas would be syphoned out of the space. Assistants called *Leichenbrenner* (corpse burners) then removed the bodies and cremated them in an adjacent crematorium. A victim's family would later receive a bogus death notice with a false statement of death citing natural causes, such as appendicitis or pneumonia. Families often received an urn with non-specific ashes that were taken from the crematoria in the killing centres.[82]

FIGURE 9 Patients at the 'Schloß Bruckberg' care institution
The Schloß Bruckberg patients being loaded aboard a 'Gekrat' bus to be transported to a killing centre, most likely Bernburg as the photo is dated early 1941.

From January 1940 to August 1941, more than 70,000 people were killed with carbon monoxide in the gas chambers of the six *Aktion T4* killing centres. Most of the people murdered in this phase of the *Krankenmorde* suffered from schizophrenia. Of those victims, most were women. Compared to male victims, the fate of women referred to the T4 committee was more influenced by their reduced capacity for work. Among those murdered in the *Krankenmorde* were people who had neither physical nor psychiatric illnesses or disabilities but who were either convicted felons, alcoholic or drug addicted, 'work shy' or classified as 'antisocial' or 'psychopath'. When it came to children, it was their perceived 'ability to learn' that primarily determined their survival.[83]

FIGURE 10 The *Aktion* T4 killing centres in Nazi Germany and Austria 1939–1941 in their present day locations

Emil B[84] was a 66-year-old married man who had lived in Stuttgart with his wife and family. We know little of his personal or family life other than he worked as a 'piano technician' with no previous health or legal problems until the period just prior to his admission to hospital. From early 1939 Emil began to suffer significant deterioration in his mental health and in late

August 1939 he was taken by ambulance to the Bürgerhospital in Stuttgart, complaining of worsening memory loss, speech problems and increasing agitation. He had likely developed *cerebral sklerosis* (vascular dementia) from a series of strokes. Due to bed and funding shortages at the Bürgerhospital, Emil was made a ward of the state in mid-September 1939 and admitted to the Christophsbad asylum in Göppingen, near Stuttgart. On arrival at Christophsbad, Emil was disorientated, agitated and suspicious. His clinical file indicates he was frequently incontinent of urine and tended to wander the wards at night; he was soon transferred to the 'agitated ward'. Emil's doctor noted that his answers to questions were limited to one or two words and tended to be repetitive or perseverative. When put to work in the hospital's workshop, Emil proved incapable of productive activity and often urinated under the workbench. Clinical entries by nursing staff indicated he could be trained only in very basic tasks—one such note uses the phrase *dressieren*—a term used when describing the training of an animal.

The last few months of Emil B's file do not record any clinical intervention or meaningful observations, perhaps indicating both that his prognosis for any form of recovery or work capacity was exceedingly poor and that his doctors had lost interest in his case. In mid-October 1940 Emil was placed on a *Gekrat* transport and transferred to the hospital at Winnethal, one of the *Zwischenanstalt* transition centres. He was murdered in the gas chamber of the Grafeneck killing centre soon after. Although not documented in his file, it is likely that Emil B's family received a bogus death certificate with a false 'natural cause' of death as was the case with thousands of other victims.

To progress and legitimate its program of elimination of the sick and disabled, the Nazi regime utilised propaganda coordinated through the *Rassenpolitisches Amt der NSDAP* (Office of Racial

Policy, the 'RPA'). The RPA produced a monthly publication *Neues Volk* (New People) which was widely distributed in the community.⁸⁵ The magazine published a series of images throughout the 1930s depicting residents of asylums and nursing homes as physically grotesque, pathetic and expensive to keep alive. One image from a 1938 edition depicts a man with cerebral palsy on a chair and supported by a strapping young *Pfleger* (male nurse). The image and caption ('60,000 Reich Marks. What this person suffering from hereditary defects costs the People's Community during his lifetime. Comrade, that is your money too'⁸⁶) poses the seemingly straightforward answer to a worrying dilemma in society. In the contest of morality and biology, it suggests, the community is best served by economic rationalism.

FIGURE 11 Propaganda image published in the journal *Neues Volk* The caption reads '60,000 Reich Marks. What this person suffering from hereditary defects costs the People's Community during his lifetime. Comrade, that is your money too.'

And yet, in reducing human life to an economic cost, the Nazis were oblivious to the implementation cost of their eradication program within the larger scope of their war endeavours.

Apart from publishing *Neues Volke*, the RPA applied numerous other strategies to enhance public support for its eugenic policies. The *Ausstellung Entarte Kunst* (Degenerate Art Exhibition) that toured Germany in 1937 had exhibited condemned works of modern art, such as those by Jewish painters Marc Chagall and Wassily Kandinksy. The RPA ordered that drawings from asylum patients be placed alongside those of the Jewish artists, intending to equate the diseased mind of the mentally ill with the 'degenerate' Jewish or Bolshevik modern artist.[87] The RPA also produced films to encourage public support for the discriminatory aims of the Nazi regime. In 1936, it commissioned the production of a silent film *Erbkrank* (Hereditary illness). The film juxtaposes grotesquely constructed images of asylum patients, often utilising light projected from below to distort appearances. It begins with a frame stating: 'What casualness and frivolity have destroyed, what thoughtlessness and lack of conscience have handed down, is protected and cared for here.' A year later, a sound-film sequel *Opfer der Vergangenheit: Die Sünde wider Blut und Rasse* (Victims of the Past: The Sin against Blood and Race) was released in cinemas across Germany. Prominent American eugenicist Harry Laughlin was so impressed with the impact of the film that he attempted to arrange for an English subtitled version to be released in the United States, although there is no record of any public screenings of the film there.[88]

In 1941, by which time nearly 70,000 adults and children with disabilities had been murdered, a final propaganda movie on the euthanasia topic, *Ich Klage An* (I accuse), was released in German cinemas. Karl Brandt, who was to become one of the Nazi officials most responsible for the *Krankenmorde*, had suggested to Propaganda Minister Joseph Goebbels that a more sophisticated message needed to be communicated to revive community support for state implemented 'euthanasia'. Goebbels worked closely with a mainstream film director, Wolfgang Liebeneiner, who had been a member of the Prussia State Theatre and later the artistic director of the German Film Academy in Babelsberg. Due to a growing public awareness of, and resistance to, the 'euthanasia program', the Nazi regime now changed its approach. Through the release of *Ich Klage An*, the Nazi regime tried to manufacture consent for the mass-murder of the sick and disabled by carefully positioning the morality of the murder process with an end-of-life decision-making dilemma in the case of a terminal illness. The screenplay for *Ich Klage An* was an adaptation of the novel *Mission and Conscience* by Helmut Unger, a Berlin ophthalmologist who later worked in the 'euthanasia' program selecting patients for death. Produced by Tobis Films, *Ich Klage An*'s plot revolves around the story of a neuroscientist's wife who experiences the onset of a rapidly progressive form of multiple sclerosis. Fearing she will suffer an undignified death she begs a family friend, who is also a physician, to administer a fatal overdose of medication. Constrained by his sense of Hippocratic duty to do no harm, the doctor resists, leaving the women's husband to administer a lethal cocktail to her in a scene that depicts the killing as a tender act of love. When later brought before court, the woman's husband passionately defends his actions as a 'mercy killing'. His pleading that it is unnatural to prolong life in all circumstances, and that it is a dutiful yet compassionate act to end suffering through a mercy death, were crafted to support the Nazi regime's argument for a lawful form of euthanasia to be a state-administered process.[89]

According to reports from the *Sicherheitsdienst* ('SD'; the intelligence and security service), *Ich Klage An* was 'enthusiastically and well received', even by some within the Protestant and Catholic churches.[90] Many who saw the film at the time reportedly responded in favour of euthanasia if it was offered as a solution to chronic suffering. From general responses to the SD's questions, the major substantive concern expressed by viewers was unease at the prospect of legalised euthanasia becoming an administrative or legal decision rather than a medical one. For the most part, German audiences did not seem to readily connect or agree with the central narrative of *Ich Klage An* and the pleadings for a state-supervised 'euthanasia' program. Where they did, such as in audiences from cinemas in Münster and Passau, the film was regarded poorly.[91]

While there may have been some sympathy in broader German society for the idea of '*Gnadentod*' (mercy death), there was resistance to a state-administered program of mass killing of disabled citizens in some sectors of the community and parts of the medical profession. The historiography of the period has traditionally highlighted the central role played by the Catholic Church in Germany in resistance to the *Krankenmorde*. But the story is much more complex.[92]

The Catholic Church's contemporary account of its role in opposing the *Aktion* T4 program suggests it led widespread resistance and expressed resolute public indignation. While many of those who died in the *Krankenmorde* came from Catholic-run institutions in Germany, it has been argued that many further potential victims were saved by the intercession of Catholic clergy.[93] Numerous senior Catholic figures are reported to have made impassioned ethical and legal based arguments against the killings.[94] In December 1940, Pope Pius XII spoke out specifically against the killing of the disabled.[95] His 1943 Encyclical *Mystici corporis Christi* declared that 'conscious of the obligations of Our high office We deem it necessary to reiterate this grave statement today, when to Our profound grief We see at times the deformed, the insane, and those suffering from hereditary disease deprived of their lives, as though they were a useless burden to Society'.[96]

The Catholic and Protestant churches in Germany were the main non-state providers of care for the sick and disabled, operating many asylums and care institutions. The 'Inner Mission' was the national Protestant health and welfare organisation, operating more than 500 care institutions, although these did not tend to provide active treatment or any kind of clinical intervention to assist its mentally ill charges. Opposition to the *Krankenmorde* in the Protestant church was, at most, sporadic.[97] The relevant Catholic organisations in Germany came under the 'Caritas' umbrella and in general were more active in resisting the Nazi's plans.[98]

From the beginning, the Nazi regime was wary of the potential responses of the main Christian churches to their program of mass killing of the sick and disabled. In January 1939 KdF bureaucrat Viktor Brack[99], whose main responsibility would be the organisation of the murder of patients, commissioned a report from Joseph Mayer, a professor of Moral Theology at the University of Paderborn, seeking advice on the likely responses of the Catholic church to the 'euthanasia' program. Mayer opined that in the event of a state-controlled 'euthanasia' program, the response of both the Catholic and Protestant churches would likely be one of acceptance if it were presented as being in the national interest. According to historian Michael Burleigh, some of the leadership of the relevant Catholic institutions entered into negotiations with T4 officials over procedural issues such as administration of last rites to the prospective victims and whether victim's ashes could be offered Christian burial.[100] It is probable that these kinds of responses emboldened Hitler in his decision to proceed with the implementation of the *Krankenmorde*.[101]

Despite initial indications of potential acquiescence by local Christian churches to the Nazi regimes' plans for the nation's sick and disabled, by late summer 1941 protests from senior Catholic clergy had caused concern for the KdF by creating unrest and resistance in numerous German communities. In March 1941 the Bishop of Berlin, Konrad von Preysing, spoke out against the killing.[102] In the months following, the words of the Bishop of Münster, Clemens von Galen, would echo within and beyond Germany and lead to him being characterised as the true gadfly of the *Krankenmorde*.

Clemens August Graf von Galen was one of 13 children of Count Ferdinand Herbert von Galen of Westphalia. Clemens was Jesuit educated as a high school student and attended the Catholic University at Freiburg in Germany's south west. After a visit to Rome in 1896 he decided to become a priest. He entered the seminary in Münster and was ordained in 1904, spending two decades in Berlin and volunteering for military service in the 1914-1918 war. Like many German ex-servicemen in the 1920s, von Galen seemed to endorse the 'stab in the back' legend and was horrified at both the collapse of the Kaiser's empire and the advent of political extremism in Germany in 1919. Von Galen had always identified as politically conservative and fiercely anti-Bolshevik.[103] He was consecrated Bishop of Münster in 1933.

Von Galen had a complicated relationship with the Nazi regime. He had preached against Nazi racial policy, although his statements were usually more a critique of the vacuous and intellectually dishonest rhetoric of the regime's racial ideologues rather than a specific repudiation of the ideas of racial hygiene and anti-Semitism.[104] In 1933, an agreement or *Konkordat* between the Holy See and the German government guaranteed that both parties would not interfere in the affairs of the other. As the Nazi regime became more totalitarian, the 1933 *Konkordat* started to break down, leading von Galen to directly assist Pope Pius XI in drafting the 1937 anti-Nazi Papal Encyclical *Mit brennender*

Sorge (With burning concern). Unlike other Encyclicals written in Latin, *Mit brennender Sorge* was written in German and smuggled into Germany. In it the Vatican condemned the Nazi-inspired 'pantheistic confusion', the 'neopagan myth of race and blood', the idolising of the State and the 'cult of the Führer'.[105] Yet, despite his criticism of many aspects of the Nazi project, Clemens von Galen remained a patriot and supported German military action against Poland and the Soviet Union.

Von Galen is most famous for his three sermons in July and August 1941 speaking out against the apparent excesses of the Nazi regime. The third—directly addressing the T4 killings[106]—was delivered in Münster's Lambertkirche on 3 August 1941 and is frequently cited as the example of the resistance of the Catholic Church to the *Krankenmorde*. Von Galen was likely inspired when he learned that patients from a Catholic-run asylum in Marienthal—part of his diocese—were to be sent to their deaths. He began this famous sermon by quoting *Luke* 19:41-44 and provided the congregation with his interpretation of the passage: that, in this instance, man was wrong in imposing his will over God. He declaimed: 'these are people, our brothers and sisters; maybe their life is unproductive, but productivity is not a justification for killing'.[107] Von Galen's sermon became inspirational to opponents of the Nazi regime and he was dubbed 'the Lion of Münster'. The text of the sermon was cited on BBC radio broadcasts to occupied Europe and dropped by the Royal Air Force as a propaganda leaflet amongst German troops.[108]

It is also likely that the sermon influenced the Scholl siblings, Sophie and Hans, and their 'White Rose' anti-Nazi movement. Founded by a group of medical students in Munich in 1942, their non-violent resistance included spreading leaflets and graffiti to condemn Hitler and the atrocities of the Nazi government. Sophie,

Hans and a fellow student activist, Christoph Probst, were later caught and quickly tried, and then beheaded in February 1943.[109] Von Galen was more fortunate—his high public profile protected him against retribution by the regime, although Hitler had said that he would exact revenge after the war.[110]

FIGURE 12 Archbishop of Münster Clemens von Galen

Other, less well-known sources of Catholic resistance paid a severe penalty. Bernhard Lichtenberg, the provost of St. Hedwig's Cathedral in Berlin, used von Galen's sermon as the pretext to send a letter of protest to Leonardo Conti, the *Reichsgesundheitsführer* (state health chief), on 26 August 1941. Citing the case of a mother who had told him of the sudden death of her son in a psychiatric institution, he demanded: 'as a human being, a priest and a German that you answer for the crimes that have been perpetrated at your bidding or with your consent'.[111] Lichtenberg not only denounced the murder of patients but also stood up for all those persecuted by the regime. He was arrested and sentenced to prison. After serving his sentence, the Gestapo had him transferred to a concentration camp. Sick and weakened by this time, Lichtenberg died in transit in 1943.[112]

The actions of Judge Lothar Kreyssig provide one of the known examples of open resistance to the *Krankenmorde* by a member of the legal profession.[113] Kreyssig had been appointed a judge in Chemnitz, Germany, in 1928. After the National Socialists took government in 1933 he refused to join the Nazi Party, preferring to join the anti-Nazi Confessing Church. In 1937 he took up a judgeship in the lower district court of Brandenburg an der Havel, including primary responsibility for mental health guardianship and the welfare of disabled adults and children. Kreyssig became concerned at the exponential growth in death certificates of these citizens and escalated his concerns to Franz Gürtner, the Reich Minister of Justice. He then learnt of the T4 program and spoke out against its illegality. In a brave and optimistic ploy, Kreyssig filed homicide charges against the KdF head Philipp Bouhler, and filed further injunctions against the transfer of patients from institutions to various killing centres. Kreyssig argued that Hitler's 1939 letter to Brandt and Bouhler was not law. Rather than withdraw his legal claims, Kreyssig took an early retirement in 1942.

There are also accounts of psychiatrists and senior medical officers resisting the murder or forced sterilisation of their patients. Karsten Jasperson was the clinical director of the Catholic-run Bethel-Sarepta asylum in Bielefeld in the north west of Germany. Jasperson refused to cooperate with local authorities in the T4 program and incited his colleagues to either not participate in the registration of patients or to alter their clinical records to make their prognosis seem more favourable and prevent their selection for death.[114] In other situations, psychiatrists may have inadvertently saved patients by misunderstanding the purpose of the T4 registration forms, believing instead they were part of a process of separating patients along clinical lines and therefore not assessing them for the gas chamber or, later, the Luminal syringe.[115] In 1940 Jasperson instituted legal action against the T4 program and is credited as the one who prompted von Galen's concern and sermons by alerting him to the pending transport of patients from Marienthal to the killing centres.

Karl Bonhoeffer, professor of psychiatry at Charité Hospital in Berlin, was the father of one of the best known figures among Protestant resistors to Nazism, Dietrich Bonhoeffer. Karl Bonhoeffer is known to have employed and protected Jewish colleagues and patients.[116] While initially supporting the sterilisation programs, Karl Bonhoeffer would later modify and reverse his position.[117] His resistance included advising his son in his efforts to try to save the lives of asylum patients—among the many other anti-Nazi activities that led to Dietrich's execution in 1945—as well as providing valuable moral and professional support to his colleagues as they resisted the T4 program.[118]

The director of Polyclinics at the Berlin Institute for Psychological Research and Psychotherapy, psychiatrist John Karl Friedrich Rittmeister, is a celebrated martyr of the Nazi period.[119] Rittmeister had studied with influential psychoanalyst Carl Jung at the Burghölzli clinic in Zurich but fell out with him over Jung's apparent allegiance to Hitler and his anti-Semitism. As a psychotherapist, Rittmeister was at odds with the views of many of his colleagues who saw mental illness as a biological phenomenon. Rittmeister is known to have hidden Jewish neighbours in Berlin after the *Kristallnacht* pogroms in November 1938 and later helped many Jewish Germans flee the Nazi regime. He is believed to have assisted directly with saving patients from the T4 gas chambers.[120] Rittmeister was an acquaintance of Harro and Libertas Schulze-

Boysen, organisers of the anti-Nazi resistance group the '*Rote Kapelle*' (Red Music Ensemble). In 1941 Rittmeister joined the organisation and co-authored 'AGIS leaflets' (named after the Spartan king) condemning the atrocities of the regime. The Schulze-Boysen's were arrested and executed in 1942, Rittmeister was executed in 1943.[121]

FIGURE 13 Karl Bonhoeffer, professor of psychiatry at Charité Hospital in Berlin
Initially a supporter of eugenic sterilisation, Karl Bonhoeffer later revised his opinion after the commencement of the T4 program. He became one of the resistors to the persecution of the sick and disabled by the Nazi regime.

Psychiatrist Walter Creutz was the chief medical officer in the administration of the Rhine Province. In January 1941 he wrote a memorandum against the 'euthanasia' measures and submitted it to the local governor, Heinz Haake. In March 1941 Creutz convened a meeting to coordinate efforts to subvert the transport of patients to killing centres which likely helped save the lives of 3,000 patients. Much of Creutz's activities against the T4 program only emerged in evidence presented to the Düsseldorf 'euthanasia trial' (1948-50) resulting in Creutz being acquitted of complicity to murder.[122]

Gottfried Ewald, director of the State Hospital and Nursing Home and the University Clinic for Psychiatry in Göttingen, refused outright to cooperate with the T4 program. He helped save potential victims by arranging urgent discharges from asylums, referring patients to other institutions and altering patient files. In mid-1940 Ewald wrote a comprehensive critique against the *Krankenmorde*, which he sent to the T4 medical office.[123]

The 1941 diary of Ernst Arlt, an Austrian psychiatrist working at the Feldhof hospital in Vienna, provides an insight into the motivations and resistance of some psychiatrists to the T4 program. Arlt's musings begin with a series of statements indicating his anger at the murder of patients and his profound disdain for the 'pseudoscience' of eugenics.[124] He notes that some of the great Germans who were afflicted by mental illness, such as Friedrich Nietzsche, would likely not have survived the program. He criticises the argument of 'incurability' as being deeply flawed, as it is possible that in the future there may be treatment that may in fact 'cure'. He documented his numerous representations to the Feldhof's clinical director, disagreeing with the transportation of patients to the killing centre at Hartheim, and records the instances he contacted families and advised them to take their relatives home to prevent them being killed. This practice, he noted, led to his being banned from using the telephone in the clinic.

In other circumstances, psychiatrists seemed compelled into a form of 'plea bargaining'[125] where the lives of some patients were sacrificed to save others. In one case, Leipzig Professor Gerhard Schorsch had compromised with T4 officials by designing a spectrum of productivity ranging from 'vegetative existences' to 'very good performance', leading to the salvation of some of those condemned by the T4 committee's assessors but at the expense of other patients, sacrificed to satisfy the local health authorities.[126]

In the eyes of the Holocaust scholar Michael Berenbaum, this kind of process created a moral equivalence between these physicians and the *Judenrat* (Jewish leadership) in the ghettos who 'horse traded' the lives of some Jews to save others from deportation to extermination camps.[127]

Despite the efforts of the Nazi regime to disguise the T4 program, the German population became increasingly aware of what was happening. The Dresden academic Victor Klemperer, whose diaries are an invaluable account of life in Germany under the Weimar and Nazi years, wrote on 22 August 1941, 'Frau Paul… talks in despair about her mother, 89, who is showing signs of senile dementia…I cannot put her in a hospital, she'll be killed there. There is widespread talk now of the killing of the mentally ill in the asylums'.[128] The SD reported that many patients refused to attend health appointments out of fear of being killed by their doctors.[129]

American journalist William L Shirer released accounts of the 'mercy killings' of Germans with disabilities in periodicals such as *Life* and *Reader's Digest* and through his journal *Berlin Diary*, all published in 1941.[130] Diplomatic reports of the murders also reached the US government before its entry into the war in December 1941.[131]

The SD also noted an incident in the Bavarian town of Absberg in which citizens had resisted the deportation to the gas chambers of disabled residents from the Ottilien institution.[132] There is an— admittedly disputed—account of Hitler himself being assailed by a crowd at the railway station in the town of Hof, also in Bavaria. Hitler's private train had stopped there whilst en-route from Munich to Berlin at the same time as a group of condemned asylum patients were being loaded onto a train to be transported to their deaths. Disturbed by what they were witnessing, and realising the arrival of Hitler's entourage, an angry crowd is reported to have gathered around the train and jeered the Führer.[133]

The T4 killing centre at Grafeneck was closed after reports of protest had reached Viktor Brack of the KdF. On 25 November 1940 an aristocrat, Else von Löwis of Menar, herself a committed Nazi and a leader of the women's Nazi movement, wrote to the wife of a presiding judge of the Nazi party court, 'Undoubtedly you know about the measures currently used by us to dispose of incurable mental patients; still, perhaps you do not fully realize the manner and the scope of this, nor the horror it creates in the people's minds! Here in Württemberg the tragedy takes place in Grafeneck, on the Alb, as a result of which the name of that place has taken on an ominous meaning'.[134] Himmler received a copy of the letter in December 1940 and wrote to Viktor Brack, 'I hear there is great excitement on the Alb due to the institution Grafeneck. The population recognises the grey automobile of the SS, and think they know what is going on at [sic] the constant smoking of the crematory. What happens there is a secret, and yet is no longer one. Thus the worst feeling has arisen there and in my opinion there remains only one thing, to discontinue the use of the institution in this place'.[135]

The sense of disquiet and increasing rumours within the community presented a problem for Brack, who then ordered the cessation of operations at both Grafeneck and Brandenburg killing centres. Not deterred, the T4 organisation commissioned two more killing centres at Hadamar and Bernburg to become operational by late 1940. Any patients from Grafeneck's previous catchment were sent to their deaths at the already-operating Hartheim killing centre.

FIGURE 14 The gas chamber at Hadamar 'euthanasia' memorial (present day)
The memorials at Hadamar and Bernburg are the only sites with gas chambers still preserved.

After murdering more than 70,000 people, the Nazi regime halted *Aktion* T4 at the end of August 1941. While the regime was troubled by the sporadic protests of the Catholic church and others, a more significant concern was public unrest at a time when Germany was escalating the war against the Soviet Union. However, the end of *Aktion* T4 would not mean the end of the *Krankenmorde*, as the intentional murder of patients would be continued by other means and to a much larger extent.

The clandestine killing process now entered a phase that has come to be known as '*dezentrale Anstaltstötungen*' ('decentralised' euthanasia). To conceal this new form of murder, the method was changed where, instead of carbon monoxide, lethal doses of medicine combined with starvation were used to kill patients. Now the responsibility of medical and nursing staff, the *Krankenmorde* took place in more than 30 different state hospitals, including Am Steinhof, Eichberg, Eglfing-Haar, Großschweidnitz, Hadamar, Kalmenhof, Irsee, Meseritz-Obrawalde and Gauheilanstalt

Tiegenhof.[136] At the same time the scope of those to be murdered was expanded to include the frail and elderly, forced labourers who had fallen ill, and soldiers suffering from disabling illness or combat injuries. These would soon be among the nearly 87,000 victims of decentralised 'euthanasia'.

At the hospital in Kaufbeuren-Irsee in Bavaria, psychiatrist Valentin Falthauser developed a diet essentially free of any nutritional value (called the 'E-Kost'[137]) to be fed to patients who had no work capacity and were condemned to die.[138] At Eglfing-Haar hospital near Munich, the clinical director and psychiatrist Hermann Pfannmüller established 'hunger houses' in buildings 22 and 25 of the institution. Patients selected for death were fed decreasing amounts of food in a diet that was deficient in protein and fat. Pfannmüller insisted that the starving patients were weighed on a regular basis and he was vigilant in preventing staff sneaking fat or protein into the condemned patient's rations. More than 400 people died in Pfannmüller's 'hunger houses'.[139] In the many patient files from Eglfing-Haar from that time (now stored in the main Bavarian State Archive in Munich) Pfannmüller's red pencil notations usually indicated the person had died by starvation. Pfannmüller would later provide evidence for the defence in the Nuremberg doctors' trial. He was arrested by US forces in 1948, convicted, and in November 1949 was sentenced to six years' prison for his role in the 'euthanasia' deaths at Eglfing-Haar.

Meseritz-Oberwalde, in what is now western Poland, became one of the most prolific killing institutions, where approximately 10,000 people were murdered. Patients were selected for death not only on grounds of work incapacity but also for disobeying hospital rules or trying to escape. It is estimated that 97 per cent of patients sent to Meseritz-Oberwalde were murdered there.[140] The institution's medical director, Walter Grabowski, implemented harsh working conditions for nursing staff which seemed to add to the brutality of their behaviour towards the patients. Grabowski's fate is unknown, although he likely committed suicide in 1945.[141] Victims of Meseritz-Oberwalde also came from other European countries and at least one US citizen died there. The institution's head nurse, Amanda Ratajczak, admitted to killing more than 1,500 people with overdoses of morphine and scopolamine. She was tried and executed by the Soviets in May 1945.

FIGURE 15 Installation from the exhibition at Psychiatrie Museum kbo-Isar-Amper-Klinikim (Eglfing-Haar) depicting the portions of starvation rations
Patients murdered in this fashion were starved to death over several weeks by progressively reducing calories per meal.

The mass-murder of disabled children by Luminal injection in *Kinderfachabteilungen* (childrens' wards) continued until the end of the war. As will be examined in the following chapters, by late 1944 more than 10,000 people, mainly sick and weak prisoners from concentration camps, would also be murdered by carbon monoxide gas in former T4 killing centres in what was termed *Sonderbehandlung* (special treatment) 14f13. Throughout the war—to free up hospital beds—more than 80,000 psychiatric patients were murdered by the SS or Wehrmacht or were starved to death in hospitals in the occupied territories.

From the time of Hitler's 1939 order authorising certain doctors to decide who among the sick and disabled were to die, it is estimated that 300,000 people were killed in the *Krankenmorde*.[142]

CHAPTER 4
'DR SCHNEIDER'

In her encounter at the Brandenburg killing centre, Elvira Hempel specifically recalled being roughly handled by nursing staff and assessed by men she presumed were doctors in the small room adjoining the gas chamber. It is highly probable that the man who turned her away that day was the 'euthanasia physician' Dr Irmfried Eberl. Eberl was medical director of the Brandenburg killing centre and used the alias 'Dr Schneider' when performing his T4 duties there. The large man Elvira hid behind was most likely Dr Heinrich Bunke, Eberl's assistant. Bunke had commenced his duties at the Brandenburg killing centre in August 1940 and, like Eberl, used a pseudonym, 'Dr Rieper'. The other men and women Elvira saw that day cannot with any certainty be identified, but they were suspected to be nurses who worked at the Brandenburg killing centre.[143]

Both Eberl and Bunke were young medical graduates who had volunteered for the job of 'euthanasia physician'. The motivations of the physicians who volunteered to work in this program varied, although at some level all were drawn to the task by the prospect of participating in a clandestine campaign carried out on the personal order of Adolf Hitler, as well as the opportunity to work with some of the most influential medical figures in Germany.[144] In Eberl's case, financial considerations will have also played a part in his choice. Eberl earned 1,000 Reichsmarks (RM) per month as director of the Brandenburg 'euthanasia centre'—paid for by the fake organisation 'Charitable Foundation for Institutional Care', one of four such 'front organisations'.[145] Eberl supplemented his income by RM 900 per month moonlighting as an occupational physician with several local companies, including the armoured vehicles manufacturer Altmärkisches Kettenwerke, the Edeka Association of German Retail Co-operatives, and the Werner & Co. publishing company. All told, his monthly earnings were almost double the typical salary for a medical practitioner at the time.[146]

Irmfried Eberl was born in 1910 in the west Austrian town of Bregenz, one of three sons of a commercial inspector. While the Eberls were Catholic, their enthusiasm for the ideas of Georg Ritter von Schonerer, the leader of the German nationalist movement in Austria, placed them at odds with the Austrian Catholic church. The Eberls converted to Protestantism in the 1930s. Both Eberl's father and older brother were among the first members of the Austrian Nazi party.

Eberl commenced his medical studies at the University in Innsbruck in 1928 and in 1931 joined the Austrian Nazi party, becoming its representative in the university's Student's Chamber. He graduated from medical studies in 1935; however his membership of the Austrian Nazi Party led to his medical license being rescinded in 1936, as the Austrian Nazi party and all its affiliated organisations had been outlawed after the assassination of Austrian Chancellor Engelbert Dollfuss by Nazi activists in

1934. Eberl fled to Germany in 1936 as 'political refugee', where he initially held various temporary positions as a doctor and later found permanent employment at the Reich Main Health Office in Berlin.

Apart from his moustache, which he altered from full and bushy to a small trimmed square like his beloved Führer, Eberl cut an unremarkable figure. In 1938 he married Ruth Rehm, a fellow medical practitioner. Rehm's family were all dedicated Nazis and were both wealthy and politically connected. Rehm often informed the Gestapo of the suspicious activities of her neighbours or acquaintances.[147] She was a close friend of the wife of Reich Health Leader Leonard Conti and in late 1939 she helped secure Eberl employment with the T4 program.[148]

FIGURE 16   Irmfried Eberl (1942)
Irmfried Eberl wearing an Organisation Todt uniform in the Minsk area in early 1942. When working as a 'euthanasia physician', Eberl used the alias 'Dr Schneider'.

After participating in a successful 'test gassing' that murdered around a dozen victims at the former Brandenburg prison complex in early January 1940, Eberl was appointed director of the new killing institution established there later that month. Beyond his duties at Brandenburg killing centre, Eberl helped devise both the T4 registration forms and develop the selection criteria for suitability for 'euthanasia'.[149] He and his associates became adept at their task of mass killing. He kept a detailed personal record of his time at both Brandenburg and later Bernburg killing centres. Eberl's diary of death listed the details of arriving transports, the institution the patients had come from, the number of victims, as well as whether the patients were women, men or Jewish. Despite the earlier proclamation of Viktor Brack, head of the KdF and *Aktion* T4, that Jews were not to be granted the 'good fortune of *Gnadentod*' (euthanasia)[150], it is now clear that Jewish asylum patients were among the first victims of *Aktion* T4.[151]

In traditional historiography, the 'euthanasia' program *Aktion* T4 is interpreted as a precondition of the Holocaust. Based on Henry Friedlander's landmark work on the origins of Nazi genocide, the main link between the 'euthanasia' program and the 'Final Solution' is seen in the murder of the Jewish psychiatric patients during the early phases of the *Krankenmorde*. At the beginning of *Aktion* T4, Jewish patients in the German Reich were systematically documented and later separated from other victims in 'collection institutions'. From the summer of 1940 onwards, all Jewish patients in asylums were killed on the sole criterion of being Jewish, regardless of their medical prognosis or ability to work. The first systematic mass murder of German Jews under the Nazi regime was perpetrated by Irmfried Eberl and colleagues in the gas chamber of the T4 killing centre in Brandenburg an der Havel.

In July 1940 every patient in asylums or other institutions in Berlin and the surrounding suburb of Brandenburg who were considered *Volljuden* ('full Jews') were sent to the Berlin-Buch State Hospital.

They were later misleadingly reported as being moved to an external institution. Witness statements and the personal notes of Eberl demonstrate that these people were transported to the killing centre in Brandenburg and murdered in the gas chamber. As best can be established at the time of writing, more than 400 Jewish patients from Berlin and Brandenburg were murdered in the T4 killing centre there. A month later, further 'collection institutions' for Jewish patients in hospitals and nursing homes in Germany were established at Langenhorn (Hamburg), Eglfing-Haar (Bavaria), Am Steinhof (Vienna), Gießen (Hesse), and Wunstorf (Hanover).

Jewish patients were held for only a few days before being transported by *Gekrat* buses to be gassed. Four hundred and twenty-five Jewish patients from the 'collection institutions' of Hamburg, Wunstorf and Gießen were murdered in the gas chamber in Brandenburg killing centre.[152] The families of these victims were then informed that their relatives had been relocated to an institution in the Polish city of Chelm. Even after they had been killed, their families received bills for treatment and care from the bogus institution, written on fake letterhead.[153] In the 'Protectorate' of Bohemia-Moravia (Czechoslovakia) Jewish patients in psychiatric hospitals and care institutions were sent to selected 'collection institutions' and then deported to the ghetto at Theresientstadt, or later to 'extermination camps' in Poland.[154]

―――

Sonia Wechsler (née Krupnick) was born in 1886 in Novoaleksandrowsk (Lithuania) near the Latvian border. She was the daughter of the Jewish estate administrator Feibusch Krupnick. As a young woman, Sonia was sent to Liepaja (Libau) in Latvia to learn tailoring. In 1907 she joined the *Bund*, a Jewish workers' association, where she met her future husband Tuvia. Sonia and Tuvia were married in 1911, and between 1912 and 1922 their four children Yaacov, Avraham, Esther and Meir were born in Liepaja. Tuvia, who was highly talented in mathematics, earned the family's income by working as an accountant for several small traders. In the winter of 1918-19 he spent a semester studying mathematics at Tübingen University in Germany. After their son Meir was born in 1922, they agreed that Tuvia should continue his studies in Germany. Taking his older sons Yaacov and Avraham with him, Tuvia returned to Germany, while Sonia stayed in Liepaja with their daughter Esther and the new born Meir. During this time, Sonia earned a living as a dressmaker and by running a small grocery shop. In 1923 she and the two younger children left Liepaja to join Tuvia and the boys in Germany.

The family settled in Hamburg, first renting a small room. It was only in 1927 that the family could move to their own apartment. Tuvia continued his studies in Hamburg while also assisting local retailers with accounting and any other job he could find. Sonia continued her contribution to the family's income through needlework.

After his father died in 1926, Tuvia's life changed profoundly. Tuvia had lived as an atheist with socialist leanings but in his grief he became deeply religious and devoted himself to the Jewish traditions of his parents. Tuvia's attempts to convert his family to Judaism caused conflict, especially with his oldest son Yaacov, an artistically talented young man who aspired to be a painter. When Yaacov was accepted at the Hamburg academy of fine arts in 1930, the dispute with his father escalated. The discontent within the family also placed Sonia under huge emotional pressure. By 1930 she was frequently receiving psychiatric treatment and in 1934 she was admitted to Friedrichsberg state psychiatric hospital.

Aware of the anti-Semitic agenda of the new Nazi regime and the increasing persecution of Jewish citizens in Germany, Tuvia sought to move his family to 'Eretz Israel' (the land Israel – then under British control as the Mandate of Palestine), and arranged the necessary papers. Like many international sites for German Jewish people seeking asylum from the Nazis, British Palestine did not permit entry of immigrants with mental illnesses. Regardless, the Wechsler family fled Germany in several phases over 1934 and 1935 but without Sonia, who remained under psychiatric care. On 23 September 1940 she was sent from Langenhorn State Hospital to the killing centre in Brandenburg an der Havel where she was murdered in the gas chamber, one of at least 2,000 Jewish victims of *Aktion* T4.[155]

After the cessation of *Aktion* T4 in 1941, Jewish psychiatric patients were persecuted and murdered in the same manner as other Jewish victims of the Holocaust. In early 1941, psychiatrist Friedrich Hetzer of Mariahilferstraße 107 in Vienna, completed a T4 registration form for one of his patients, a 48-year-old married woman diagnosed with schizophrenia. Hetzer reported that his patient had grandiose ideas and heard voices. Her medical file documented that she had written letters to heads of governments advising of her views on the conduct of the war. She was occasionally capable of work in the sewing room. The registration form noted that the patient's sister was mentally ill, her brother had committed suicide, her mother was an 'hysteric' and her father was 'neurasthenic'.[156]

Dr Hetzer's patient was Margarethe Neumann (née Herzl), the youngest daughter of Theodore Herzl, the father of Zionism.[157] Known to her family as 'Trude', Margarethe had attended a gymnasium in Vienna through the generosity of Zionist benefactors. Her education was disrupted after she suffered an episode of manic psychosis age 17, and then dropped out of school. Trude attempted various university courses but was unable to complete any form of study. She was not hospitalised until her mid-20s, suffering another psychotic episode of what was proving a severe and persistent mental illness. One of her father's biographers noted Margarethe's 'almost uncanny parodies of her father's dynamic exuberance and epistolary style carried to psychotic extremes'.[158] Margarethe would spend almost a decade in private sanatoria and was certified a ward of the court. At the time of her T4 registration she was resident in the Sanitorium Purkersdorf on the southern outskirts of Vienna, and like all Jewish patients in Austrian private sanatoria, she was sent to the Am Steinhof institution in the XIV District in outer western Vienna in March 1941. In January 1942 she fell onto a marble floor while dusting a side board, fracturing her right shoulder and right neck of femur (hip) and was sent for surgery. After convalescing in an aged care facility in Lainz, she was returned to Am Steinhof in May 1942.[159] There is no information in her file as to why she received this specialised medical care for her injuries. Her experience highlights the complex and sometimes inexplicable dissonance between the brutal treatment and murder of patients in some circumstances and their receipt of special care in others.

In September 1942, Margarethe and her husband Richard were deported to the Jewish ghetto in Theresienstadt (Czechoslovakia) on transport number IV/10–967. Theresienstadt was established as a 'transit ghetto' in November 1941 for the Jewish population of the Protectorate of Bohemia and Moravia. From the summer of 1942, older Jewish deportees from Germany and Austria, like Margarethe and Richard Neumann, began arriving at the ghetto.[160] As in all Nazi imposed ghettos, Theresienstadt had a Jewish Council (*Judenrat*). Compared to the other Jewish ghettos in Nazi zones of occupation, the Theresienstadt *Judenrat* had more autonomy in day-to-day administrative decisions and this provided the Jewish leadership with more opportunities to draw on the pooled expertise of deportees with professional backgrounds to improve the dismal quality of life there. In this context, three doctors—Erich Munk (a radiologist), Erich Springer (a surgeon), and Erich Klapp (an internist)—established an elaborate health service (*Gesundheitswesen*) in the ghetto that, at its peak operation, provided one physician for every 500 people, and eight general or specialist hospitals and care institutions for elderly people. Viktor Frankl, the Viennese neurologist and psychotherapist who would later become famous for his landmark book *Man's Search for Meaning* and the establishment of 'logotherapy'[161], had arrived in Theresienstadt in September 1942. Frankl first set to work as a general practitioner, although he soon took up practice as a psychiatrist in Block B IV (the psychiatric wing of the *Gesundheitswesen*), where he established therapeutic services to help new arrivals to the ghetto adjust to their situation and reduce the suicide rate among the deportees.[162] Because the SS feared the Theresienstadt ghetto becoming a nidus for infectious disease that would spread to their personnel, they provided the physicians of Theresienstadt with reliable supplies of medications, including expensive antibiotics.[163] Yet despite the valiant efforts of the physicians and nurses in Theresienstadt, it is estimated that by the end of the war around 34,000 people (mainly elderly) died from malnutrition and its consequences.[164]

Like so many in Theresienstadt, Margarethe Neumann died in unknown circumstances in mid-March 1943. Her body was cremated along with 23 other people and her remains were never found. Richard Neumann died soon after. The Neumann's only son, Stephan—who had been sent to England in 1935—died by suicide in Washington DC in 1946, after learning the fate of his parents. Margarethe Neumann is the only member of the Herzl family not buried in the Israeli cemetery at *Har ha-Zikaron*.

FIGURE 17 The Theresienstadt 'camp-ghetto' site (present day)
The entrance to the 'small fortress' of Theresienstadt, a walled garrison town in what was then Czechoslovakia. Theresienstadt was run as a prison by the Gestapo, becoming a ghetto and concentration camp from 1941. Approximately 155,000 people were sent there by the time of its liberation in 1945. More than 34,000 people died within its walls and 88,000 were deported via Theresienstadt to extermination camps further east. After arriving in the ghetto in 1942, Margarethe (née Herzl) Neumann died there in unknown circumstances in 1943.

In addition to military preparations for the invasions of Poland and later the Soviet Union, the Nazi regime's plan for colonisation and settlement of lands to the east involved detailed consideration of political reorganisation and annihilation of racial 'inferiors', most significantly Jewish communities. Known as *Generalplan Ost*, developed and implemented between 1940-1942, the comprehensive and intentionally secret plan called for a process of economic exploitation of Jewish civilians in occupied territories and their ultimate 'resettlement'.[165] Under Hans Frank's *Generalgouvernement* in Poland (1939-1945), a series of Jewish ghettos were established in major Polish cities while the Nazi leadership deliberated on the fate of Jews within the expanded Reich.

As part of the invasion of Poland in September 1939, special paramilitary groups were established and tasked with the murder of large numbers of Jewish people and anyone who posed a risk of political opposition to the occupation. These formations, known as *Einsatzgruppen* (operational groups), were organised as discreet units from other military groups. Seven *Einsatzgruppen* of battalion strength operated in Poland in 1939.[166] After the invasion of the Soviet Union in June 1941, *Einsatzgruppen* were reorganised into four larger units operating behind the *Wehrmacht's* lines of advance. As in their remit in Poland, their task was in the first instance to murder Jewish men, Bolshevik Commissars and other 'undesirables' or potential sources of opposition to the occupation.

Several commanders of *Einsatzgruppen* were prosecuted by the International Military Tribunal (IMT) in Nuremberg after the war. One defendant, Otto Ohlendorf, had commanded *Einsatzgruppe D* which had operated behind the advancing German 11th Army in Moldova, south Ukraine, and the Crimea. Ohlendorf's deposition outlined the organisation and *Aktionen* (operations) of *Einsatzgruppen*: 'According to an agreement with the Armed Forces High Command...the Special Commitment Detachments (*Einsatzkommandos*) within the army group were assigned to certain army corps and divisions...The unit selected for this task would enter a village or city and order the prominent Jewish citizens to call together all Jews for the purposes of resettlement. They were requested to hand over their valuables to the leaders of the unit, and shortly before their execution surrender their outer clothing. The men, women and children were led to a place of execution which in most cases was located next to a more deeply excavated anti-tank ditch.'[167]

In August 1941 *Reichsführer-SS* Heinrich Himmler visited his men operating in Byelorussia (present day Belarus). On 15 August 1941 the commander of *Einsatzgruppe B*, Arthur Nebe, arranged for Himmler to tour the ghetto at Minsk and then view a mass execution performed by his men. A group of 98 men and two women had been selected and were executed by rifle fire in front of Himmler. Himmler's adjutant, Karl Wolff, described the scene: 'An open grave had been dug and they had to jump into this and lie face downwards. And sometimes when one or two rows had already been shot, they had to lie on top of the people who had already been shot and then they were shot from the edge of the grave. And Himmler had never seen dead people before and in his curiosity he stood right up at the edge of this open grave – a sort of triangular hole – and was looking in. While he was looking in, Himmler had the deserved bad luck that from one or other of the people who had been shot in the head he got a splash of brains on his coat, and I think it also splashed into his face and he went very green and pale – he wasn't actually sick, but he was heaving and turned round and swayed and then I had to jump forward and hold him steady and then I led him away from the grave'.[168]

Later in the day Himmler visited an asylum at Noviniki near Minsk during which he instructed Nebe to arrange for the 'liquidation' of the residents. Also present was Nebe's commander, Erich von dem Bach-Zelewski. Because of his experiences that morning, Himmler insisted that they discuss the issue of killing methods and the harmful psychological effects that shooting civilians had on the men.[169] Himmler indicated that a more efficient and 'humane' method of mass murder needed to be found, particularly as the killings were soon to include women and children. After the discussion, Himmler determined that poison gas in static and mobile gas chambers should be the method used to kill the millions of intended future victims. This decision about the changed method of mass killing and the expansion of victims would become what historian Christopher Browning termed a 'caesura in the history of the Holocaust'[170], marking the end of one phase of the persecution and sporadic murder of Europe's Jewish population and the commencement of a larger scale, coordinated attempt at genocide.

From the experiences of *Einsatzgruppen* in Poland there was ample evidence that participation in mass executions by gunfire at close range caused the killers to abuse alcohol or suffer intense psychological distress. Zelewski had argued that the killing process meant that 'these men are finished for the rest of their lives' becoming 'either neurotics or savages'.[171] Up to 20 per cent of men in *Einsatzgruppen* formations exhibited a psychiatric syndrome akin to combat neurosis[172] or what would be now considered posttraumatic stress disorder. During the trial of the senior Nazi leadership at the International Military Tribunal (IMT) in 1945, Otto Ohlendorf testified as a witness. He told an American psychiatrist Leon Goldensohn[173] of having to relieve his psychiatrically disturbed men from their duties, as well as his own insomnia and disturbed appetite during his period on the Eastern Front.[174]

There is further evidence of the psychological toll of the mass killings on *Einsatzgruppen* personnel within senior ranks of the SS well before Himmler's presyncope in Minsk.[175] Zelewski himself was long troubled by gastrointestinal complaints and chronic constipation arising from morphine abuse, likely used to self-medicate anxiety emerging from his duties. He was to later suffer what appeared to be a psychotic episode in March 1942 while in Hohenlychen clinic in Berlin.[176]

FIGURE 18 *Einsatzgruppen* murdering a group of Jewish victims in Ivanagorod in Ukraine, 1942
The psychological burden of killing men, women and children by shooting at close range so concerned Heinrich Himmler that he insisted other methods for mass killing be found.

The definitive point at which the Nazi leadership decided to murder the entirety of Europe's Jewish population is still debated. It is likely that the summer of 1941 was the moment of determination: the time when other options for resolving the *Judenfrage* (Jewish Question) were excluded. Viktor Brack had proposed to Himmler that the economic benefits of putting Jewish victims to work would be squandered by mass killing, proposing that 'sterilization, as normally performed on persons with hereditary diseases, is here out of the question, because it takes too long and is too expensive. Castration by X-ray however is not only relatively cheap, but can also be performed on many thousands in the shortest time'.[177] This and the previously planned forced migration of all Jews in the Reich to Madagascar were deemed impractical by the Nazi regime.[178] In October 1936 Hitler gave *Reichsmarschall* Hermann Göring responsibility for implementing a Four-Year Plan, which included dealing with the *Judenfrage*. In January 1939 Göring authorised Reinhard Heydrich[179], the head of the *Reichssicherheitshauptamt*-RSHA (Reich Security Main Office) to establish an organisation to deal with issues of Jewish emigration and expulsion from Europe. By late July 1941 Göring instructed Heydrich to 'carry out all necessary preparations with regard to organisational, substantive, and financial viewpoints for a total solution of the Jewish question in the German sphere of influence in Europe'.[180] This task was to become known as the *Endlösung* (Final Solution).

During March 1941 Himmler met with Bouhler to discuss the possibility of deploying the T4 killers to eliminate concentration camp prisoners who, weakened through starvation and illness, could no longer be exploited as forced labourers. However, the concentration camps located in Germany and occupied Poland were not equipped with gas chambers at that time. Consequently, Himmler and Bouhler agreed that from April 1941, sick or incapacitated prisoners would be selected for death by former *Aktion* T4 physicians who were now required to inspect the concentration camps. After the 'selections', these prisoners were transported to former *Aktion* T4 killing sites that were still operational at Bernburg and Pirna-Sonnenstein in Germany, and Hartheim in Austria. The first 269 victims of the campaign termed *Sonderbehandlung* (Special treatment) 14f13 were transported in June 1941 from the Sachsenhausen concentration camp near Berlin to the killing facility at Pirna-Sonnenstein and gassed. Later transports of victims from other concentration camps such as Dachau and Mauthausen were sent to Hartheim. Irmfried Eberl and much of his original team from Brandenburg participated in this killing program, utilising the facilities at the Bernburg killing centre. Some 10,000 concentration camp prisoners from countries all over Europe would be murdered in this way before the *Aktion* 14f13 operation ceased in March 1943.[181]

The decision was taken to concentrate the proposed mass-killing of Poland's Jewish population in Lublin and the task fell to Lublin-based Odilo Globočnik, an *SS-Brigadeführer* and police chief.

With cessation of the *Aktion* T4 gassing program in August 1941, numerous personnel skilled in the art of mass killing with gas were left without a remit. Globočnik met with the KdF's Brack and Bouhler in October 1941 and had a series of meetings with Himmler to discuss the proposition of applying the *Aktion* T4 method to the mass-murder of Poland's Jews.[182] As a result, around 90 of the T4 killers were transferred to Globočnik's command. In December 1941, Globočnik ordered construction of a prototype extermination camp at Bełżec, south of Lublin near the Ukrainian border. Following Himmler's agreement, Globočnik oversaw construction of a training camp at Trawniki, a village some 40 kilometres south of Lublin, where POWs from Soviet territories and ethnic Germans from the Ukraine trained to become guards at the new mass killing camps. By 1944 more than 5,000 *Trawniki* guards had passed through the camp.[183] The SS referred to these men as *Hiwis*, an abbreviation of *Hilfswilliger* (willing to help).[184] In 1944 these troops would prove ruthless in suppressing the Warsaw uprising.

On 20 January 1942, 15 political, security and military officials met at a villa at 56–58 Am Großen in Wannsee, on the western outskirts of Berlin. The attendance list at the 'Wannsee Conference' included *SS-Obersturmbannführer* (lieutenant colonel) Adolf Eichmann, chief of the RSHA Department of Jewish Affairs. Eichmann was to become the totemic figure in the organisation and implementation of the Final Solution. The actual decision about the use of *Giftgas* (poison gas) was never documented and much of the discussions at Wannsee were transacted in euphemisms.[185]

In the aftermath of the discussions at the Wannsee meeting, other extermination camps were established in Poland in what became known as *Aktion Reinhard*.[186] In addition to Bełżec, Reinhard camps placed under Globočnik's jurisdiction were Sobibor (near Włodawa on the border of Belarus and Ukraine, on the banks of the Bug River) and Treblinka (north east of Warsaw).[187] The magnitude of the task of killing the entirety of Poland's Jewish population, and Himmler's desire to spare his men the psychological trauma of shooting innumerable civilians, meant that the demonstrated effectiveness of the *Aktion* T4 model of mass transportation and killing people in gas chambers became the preferred method.

Independent of the activities of the KdF, other groups of Nazi killers had murdered thousands of patients in asylums and care institutions in Poland and other occupied territories. Immediately following the German invasion of Poland in September 1939, adults living in psychiatric asylums and nursing homes were murdered by *Einsatzgruppen* units in sporadic massacres in the 'Warthegau' (now Western Poland). The first known massacre of Polish psychiatric patients occurred in early September 1939 in Conradstein (present day Kocborowo) in the nearby forests of Szepegawski. Later in September 1939 *Einsatzgruppen* men committed another mass killing of asylum patients in Danzig (present day Gdańsk). In October 1939 a special detachment of SS troops (*SS-Sonderkommando*) under the command of *SS-*

Untersturmführer Herbert Lange were tasked with killing thousands of psychiatric patients.[188] Lange's men murdered around 1,000 patients from the asylum at Owinksa, north of the city of Posen (present day Poznań) by pistol shot to the back of the neck.[189] Later that month Lange's men would kill 170 patients from other institutions in Posen in a makeshift carbon monoxide gas chamber constructed within a fortress known at the time as 'Fort VII', site of the newly established *Konzentrationslager Posen* (Posen concentration camp). A chemist working for the SS, Dr August Becker, conducted trials at Fort VII using both carbon monoxide and the cyanide-based poison gas Zyklon B.[190]

The further deployment of technology to accelerate mass murder included developments at the Criminal Technical Institute of the German Police, where engineers designed and trialed a prototype mobile gas chamber. This comprised a trailer pulled by either a truck or tractor, into which carbon monoxide was piped from attached pressure cylinders obtained from the factory in Ludwigshafen operated by the German chemical conglomerate IG Farben.[191] Lange's men later used mobile gas chambers (*Gaswagen*) to murder sick and disabled people. The *Gaswagen* were disguised with the logo of the well-known food store chain '*Kaiser's Kaffeegeschäft*' (Kaiser's coffee shop). The *Gaswagen* would kill the victims with carbon monoxide gas as it drove to a pre-prepared mass grave.[192] Later versions of *Gaswagen* with the gas chamber fitted onto the chassis were produced by the Gaubschat car works factory in Berlin. Mobile gas chambers that could kill 50 to 70 victims at a time were retrofitted onto various models of trucks. The truck's exhaust pipe was attached to the sealed gas chamber after the victims were loaded on board and locked in, after which the carbon monoxide-rich engine exhaust would asphyxiate them.[193] From late 1941 Lange's men murdered more than 300,000 Jewish,

Sinti and Roma people using mobile gas vans at the Chelmno extermination camp near the Polish city of Łódź.

FIGURE 19 A Gaswagen (gas van or mobile gas chamber)
This Magirus-Deutz chassis was fitted with a sealed chamber into which exhaust was pumped, killing the victims as they were driven to a burial site. This van was likely used at the Chelmno extermination camp.

In January 1942 Irmfried Eberl and around 400 former *Aktion* T4 staff members, many of them physicians or nurses, were sent to Minsk as part of the *Osteinsatz* ('medical operation in the east'), ostensibly to assist with the medical care and evacuation of wounded soldiers from the eastern front. The *Osteinsatz* was coordinated by 'Organization Todt', a special Nazi organisation responsible for military engineering projects and arms production.[194] It is now apparent that another purpose of the *Osteinsatz* was to 'euthanise' seriously wounded German soldiers, who had sustained severe brain injuries or other untreatable wounds. Pauline Kneissler, a nurse who had worked previously at the Grafeneck and Hadamar killing centres, disclosed to a confidant after the war that she had been sent to Minsk with other T4 personnel and ordered to give lethal injections to severely wounded German soldiers.[195]

Following a brief interregnum in Minsk in early 1942, Irmfried Eberl and some 90 *Aktion* T4 operatives were relocated under Globočnik's command in preparation for the mass killing of Poland's Jewish population. Eberl and his colleagues were co-opted into the SS and given honorary ranks and uniforms. Eberl first spent a brief period as an observer at the extermination camp at Sobibor. Building on the experience of mass killing using static gas chambers at Bełžec, Sobibor had been established by *SS-Hauptsturmführer* Franz Stangl and became operational in May 1942. Following this exposure, Eberl was deployed to the *Reinhard* death camp at Treblinka, north-east of Warsaw.

The Treblinka extermination facility (Treblinka II) was constructed next to an existing prison camp (Treblinka I). The site was located 4 kilometres from the village of Treblinka, with the village's main railway station in the middle of a forest near the town of Malkinia. The camp complex was near both the Warsaw-Bialystok and Malkinia-Siedlce rail lines, enabling easier transport of victims from multiple sites. As the culmination of learning from the *Reinhard* operation, Treblinka II was expected to be the most efficient of the *Reinhard* camps. It was constructed under the supervision of Erwin Lambert, who had started his career as a stone mason and had distinguished himself in the ranks of the German Labour Front.[196] The KdF recruited Lambert in 1939 to supervise the renovation of the Liebermann villa at *Tiergartenstraße* 4, and he would go on to direct the installation of the gas chambers at all the *Aktion*

T4 killing sites, his travels earning him the nickname 'the flying architect of T4'.[197]

Treblinka II comprised three sectors. 'Camp 1' housed the administrative compound and accommodation for guards and lower ranking personnel, including a small zoo set up by *Oberscharführer* (sergeant) Kurt Franz.[198] 'Camp 2', the *Auffanglager* (lower camp), was the receiving area. A railway unloading zone, comprising a long and narrow platform surrounded by a barbed-wire fence, extended from the Treblinka rail line into the camp proper. 'Camp 2' contained a building disguised as a railway station complete with a wooden clock and fake rail destination signs and timetables. Two large barracks, where victims undressed prior to entering the gas chamber, stood around 100 metres from the rail line. Next to the fake railway station building was a 'sorting square' where *Sonderkommando* ('special unit' comprising prisoners spared death to perform forced labour) collected the victims' belongings and bundled them to be sent to Germany. Next to the sorting square was the *Lazarett*, a fake infirmary disguising an execution site over a pit. Small children and victims too frail to walk to the gas chambers were taken to this site and killed by shots to the neck and their bodies dumped into the pit.

'Camp 3' was the site of the two gas chambers. Located at the southwest part of the camp, the gas chamber complex comprised a 200 square-metre area completely separated from the rest of the camp by a wall of barbed wire into which birch tree branches were woven to obscure the view. The gas chambers at 'Camp 3' were housed in a brick building in the centre of the compound where a Star of David adorned the gable. The access path to the gas chamber was called 'the tube' or what the SS referred to sarcastically as *Himmelstraße* (the road to heaven). Each gas chamber had a thick airtight door made of wooden beams 2.5

metres high and 1.8 metres wide. Like the gas chambers in the T4 killing centres, the Treblinka gas chambers were disguised as communal showers—the walls were covered with white tiles and a series of fake water pipes and shower heads installed along the ceiling. Rather than rely on externally sourced and expensive carbon monoxide, the poison gas used in the *Reinhard* gas chambers was generated from an engine, likely one removed from a captured Red Army truck or tank.

To the east of Treblinka's gas chambers were long ditches where *Sonderkommando* threw the bodies of the victims after gassing. With mass burial proving impractical due to the large number of victims, a decision was taken to burn the bodies and *Sonderkommando* were directed to pile the corpses on a grid of old railway tracks laid across a large fire pit. In addition to the *Sonderkommando*, Treblinka utilised a mixture of personnel. The main staff comprised 20 German SS troops supported by up to 120 *Hiwis*.

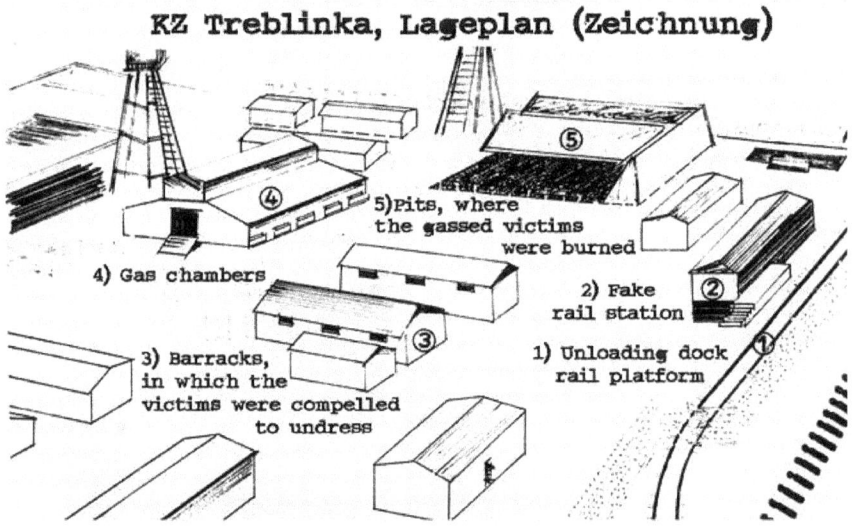

FIGURE 20 Map of the Treblinka extermination camp
The image was used in evidence in the trial of Franz Stangl.

From the outset Eberl sought to live up to expectations that Treblinka would be the most efficient of the Reinhard death factories and, for a time, he achieved the 'highest killing rate' of any site in the Holocaust: 280,000 victims in first six weeks of Treblinka's operations.[199] In July 1942 he wrote to his wife Ruth: 'I know that I have not written much to you lately, but I could not help this, since the last Warsaw weeks have been accompanied by an unbelievable agitation and likewise here in Treblinka we have reached a pace that is downright breathtaking. Even if there were four of me and each day was 100 hours long, this would surely not be enough (...) By employing myself ruthlessly, I have nevertheless managed the last days with only half of the personnel at my command. I have deployed my people ruthlessly wherever it was necessary and they have struggled along valiantly. I am happy and proud of this achievement. (...) Since you represent for me the beautiful part of my life, you should not know everything about it'.[200]

Yet despite Eberl's braggadocio, a number of his letters communicate a sense that he was out of his depth at Treblinka. It is alleged that he was frequently intoxicated and had taken to stealing the victims' belongings, as well as allowing *Hiwis* to loot the piles of clothes and luggage, often before the victims had even undressed.[201] While Eberl claimed he had coped well with both the number of victims and pace of the work of killing at Brandenburg and Bernburg, his men were overwhelmed by the uncontrolled manner of arrivals of numerous transports to Treblinka. Burial pits soon overflowed and rail carriages full of victims backed up. Locked in the rail cattle cars, victims frequently died from hyperthermia or suffocation in the summer heat. Many of the *Hiwis* operating the gas chambers were incompetent or inadequately trained, often not waiting long enough for the victims to die so that when the gas chamber doors were opened, some victims were still alive and had to be forced back in for a second gassing. The engine generating the poison gas frequently broke down, forcing terrified victims to wait in the dark gas chamber, often for hours, until it was repaired.[202] At other times, when the gas chambers were struggling to keep up with the number of victims, Eberl ordered the *Hiwis* to shoot the new arrivals on the ramp at 'Camp 2'.[203] Willi Mätzig, the so-called 'Gunman of Treblinka', whose primary task was to shoot victims behind the *Lazarett*, testified, 'When I came to Treblinka the camp commandant was a doctor named Dr Eberl. He was very ambitious. It was said that he ordered more transports than could be 'processed' in the camp. That meant that the trains had to wait outside the camp because the occupants of the previous transport had not yet all been killed. At the time it was very hot and as a result of the long wait inside the transport trains in the intense heat many people died. The *Hauptsturmführer* Christian Wirth came to Treblinka and kicked up a terrific row. And then one day Dr Eberl was no longer there…'.[204]

Christian Wirth was the General Inspector of the *Reinhard* camps. Wirth had performed a similar role in *Aktion* T4 and, prior to his elevation to a supervisory role in *Aktion Reinhard*, he had established the extermination camp at Bełżec. After reports had reached Globočnik of the situation at Treblinka, Wirth was sent to deal with Eberl. Enraged at the gross incompetence at Treblinka, Wirth terminated Eberl's command on 26 August 1942 and took over as temporary commandant.

Wirth had worked as a homicide detective in the Stuttgart *Kriminalpolizei* in the 1930s and had been an early adopter of National Socialism. Known as 'Christian the terrible', Wirth's harsh physical features and apparent coarseness accompanied his legendary ruthlessness and brutality. At various points Wirth had been a member of the SA brown shirts, the SD, and later the SS. In

1939 he was deployed to the prison at Brandenburg an der Havel where he coordinated the initial gassing experiments in January 1940. He later oversaw the development of the T4 killing centres at Grafeneck and Hartheim, where he met Franz Stangl who was later to be commandant at Sobibor.[205] Stangl told journalist Gitta Sereny after the war that: 'Wirth was a gross and florid man. My heart sank when I met him. He stayed at Hartheim for several days that time and often came back. Whenever he was there he addressed us daily at lunch. And here it was again this awful verbal crudity: when he spoke about the necessity of this euthanasia operation, he was not speaking in humane or scientific terms, the way Dr Werner at T4 had described it to me. He laughed. He spoke of 'doing away with useless mouths', and that 'sentimental slobber', about how such people made him 'puke''.[206]

After Wirth dismissed Eberl, transports to Treblinka were halted on 28 August 1942. Because Sobibor's operations had been suspended due to railway repairs, Globočnik transferred Franz Stangl to command Treblinka on 1 September 1942. Stangl told Sereny that the shambles he encountered at Treblinka was his worst experience of the Holocaust. He equated the putrefying remains of victims, overflowing burial pits, and the depredations of the *Hiwis* as 'Dante's inferno come to life'.[207] More than 763,000 people were killed at Treblinka between July 1942 and April 1943, including psychiatrist Sigmund Freud's three sisters Marie, Pauline, and Rosa on the same transport from Theresienstadt in September 1942.[208]

Eberl returned to Bernburg to reprise his *Aktion* 14f13 duties then subsequently took a prolonged leave of absence, during which time he cared for his wife Ruth who was gravely ill with a terminal illness.[209] After she died in 1944 Eberl was drafted into the *Wehrmacht* and served in Luxembourg as a medic with the 902nd Armoured Infantry Division.[210] He was captured by US troops in 1945 and worked in the POW camp's infirmary before being released from US custody in July 1945. Eberl resettled near his late wife's parents in Blaubeuren and, in order to resume practise as a doctor after the war, he applied for an accelerated handling of his denazification process. Despite his desire to hide his involvement in *Aktion* T4, *Aktion* 14f13 and *Aktion Reinhard*, Eberl used his own name in the application. He returned to medical practice and in 1947 married Gerda Poppendieck, who bore him a son.

The denazification court of Ulm later found evidence of Eberl's crimes at Bernburg and the public prosecutor's office instigated formal investigations. On 7 February 1948 American Military Police took Eberl into custody. A former nurse who had worked at Grafeneck learned of Eberl's arrest and reported her knowledge of his activities in the T4 program to police in nearby Tübingen.[211] A few days later a police officer showed Eberl a copy of a book written in 1946 by a survivor of the Buchenwald concentration camp, Eugen Kogon.[212] Kogon's book outlined in detail the activities of the SS during the war and named Eberl as a 'euthanasia doctor'. After nine days in custody, Eberl committed suicide, hanging himself in his cell.[213]

Like so many of the perpetrators of the Holocaust, Eberl was never made to explain his actions. His decision not to kill Elvira Hempel seems an outlier when compared to his later behaviour. His life reads as a study of vanity, ideological fervour and opportunism: a gross lack of empathy combined with ambition and arrogance empowered by a malignant political philosophy.

There have been many attempts by scholars to understand or at least define the motivations of the perpetrators of the medical crimes committed under the Nazi regime. Various hypotheses for their actions have appeared over time, such as the influence of politics and ideology; opportunism and careerism; institutional loyalty[214]; a servile adherence to doctrine and a distorted sense of obedience; vanity and ambition; 'inertia' of the will, and personality vulnerability.[215] Some writers have formulated perpetrator motives as seemingly self-serving variations of the concept of empathy.[216]

The motivations of members of the nursing profession who participated in the killings is also a conceptual challenge, particularly given their direct proximity to and agency in the killing process. Theories as to why they participated in the T4 program have varied between ideological or religious motivations, institutional influence, or by duress.[217] Other authors have defined the rationale of health professionals to participate in the T4 killing centres purely in terms of economic necessity: simply an issue of income supplementation.[218]

Whatever his motivation, Irmfried Eberl's roles in both the *Krankenmorde* and the later *Final Solution* embody the relationship between these profound crimes and the moral trajectory of a medical profession that, when faced with clear ideological and professional conflict, chose to support and legitimate the Nazi regime's policies and practices.

CHAPTER 5
'THERE WAS WONDERFUL MATERIAL AMONG THOSE BRAINS'

Werner Przadka was born in 1927 in Züllichau, modern day Sulechów in western Poland. At age three, Werner fell from a window and sustained a traumatic brain injury. He later suffered seizures, developmental delay and learning difficulties. At age eight he was diagnosed with a brain tumour, although an exploratory craniotomy could not confirm the location or type of lesion. Werner continued to suffer intractable epileptic seizures and in 1937 he was hospitalised in Potsdam. As his developmental and behavioural difficulties worsened, doctors declared him 'ineducable', and in 1940, at age 13, he was transferred to the *Kinderfachabteilung* (special children's ward) at Brandenburg-Görden State Hospital near Berlin. On 28 October 1940 Werner, along with more than 50 other children, was murdered in the gas chamber at the Brandenburg killing centre.[219]

According to the diary of Irmfried Eberl, Dr Julius Hallervorden—an ambitious neuropathology researcher—was present at the gassing of the children.[220] After Werner was killed, Hallervorden watched as Heinrich Bunke removed Werner's brain and placed it in fixative. In addition to Werner's brain, other tissue samples harvested from the killing were sent to Hallervorden's lab in the Kaiser Wilhelm Institute (KWI) in the Berlin suburb of Buch. Hallervorden later examined Werner Przadka's brain and found an oligodendroglioma, a type of brain tumour in the cerebral cortex, and concluded that the tumour had developed as a consequence of brain trauma from the 1927 fall. In 1948—three years after the end of the Nazi regime—Hallervorden would publish his findings in the prestigious German scientific journal *Nervenartz*.[221] The paper was cited as a reference in the scientific literature as recently as 2011.[222]

It would later be established by historians that Werner was one of at least 340 children and adolescents from the Brandenburg-Görden State Hospital who were murdered at the Brandenburg killing centre, so that their brains could be harvested for research at the KWI laboratories.

―――

Julius Hallervorden was born in 1882 in Allenberg, East Prussia. Hallervorden's father was a psychiatrist and Julius grew up in the grounds of psychiatric hospitals in Allenberg and later Königsberg. Hallervorden junior completed his schooling and medical studies in Königsberg in 1909. After graduating he pursued further training as a *Nervenarzt* (neurologist) in a private clinic in Berlin. In 1913 he gained a position in the psychiatric clinic in Landsberg an der Warthe in Prussia, modern day Gorzow in western Poland.[223] During the 1914-1918 war, Hallervorden worked in a military hospital treating soldiers suffering an array of brain and peripheral nerve injuries. As many of these patients died after a period of observation of their clinical signs and symptoms, there were ample opportunities for clinical and neuropathological research. This professional experience fostered in Hallervorden an interest in neuropathology, the branch of medicine concerned with diseases of the nervous system. After the 1914-1918 war, he worked with Walther Spielmeyer in the department of neurohistology at the German Psychiatric Research Institute in Munich. In 1924 this became part of the prestigious Kaiser Wilhelm Society for the Advancement of Science. It was during this time that Hallervorden would be influenced by the work of the biological psychiatrist Emil Kraepelin, one of the founders of modern psychiatry.

FIGURE 21 Werner Przadaka, a 13-year-old boy murdered on 28 October 1940 at the Brandenburg killing centre
Werner's brain was removed soon after he was gassed and sent to the laboratory of Julius Hallervorden in Berlin-Buch.

In 1921 Hallervorden began a professional collaboration with neuropathologist Hugo Spatz and in 1922 the pair published a case analysis of five sisters who had died after progressive dementia and speech difficulties.[224] Post-mortem examination of the brains of the young women revealed brown discolorations in different areas of the brain tissue.[225] This condition was subsequently named *Hallervorden-Spatz disease*.[226] Hallervorden's meteoric rise in academia continued and by 1929 he was head of the 'central dissection house' for the state hospitals in the Province of Brandenburg. In this position Hallervorden first developed a relationship with the young and ambitious child psychiatrist, Hans Heinze.

New opportunities arose for Spatz and Hallervorden when the Kaiser Wilhelm Institute of *Hirnforschung* (the KWIHF, a brain research institute) was founded in 1937 by highly respected neurologists Oscar and Cecile Vogt. Oskar Vogt was a clinician with dual psychiatry and neurology qualifications and Cecile was a neuropathologist.[227] They had established a neurology research institute in 1898 within the Charité Hospital in Berlin where their collaboration had integrated the studies of brain pathology and psychiatric disorder.[228] Their pioneering work had earned them the privilege of being given custody of the brain of Vladimir Lenin after he had died from a cerebral haemorrhage in 1924.[229] Hugo Spatz, then working in Munich, was appointed foundation director of the KWIHF in 1937. One of Spatz's major achievements was to establish a cooperative relationship between the KWIHF and the department of pathology at the psychiatric sanatorium in the Berlin suburb of Buch. Following implementation of the Nazi's 1933 hereditary health laws, there was a significant public and political imperative to identify and eliminate causes of heritable conditions, which in turn fostered a shift in the focus of medical research from healthy to diseased brains.[230] With numerous sites around Germany containing large numbers of newly classified 'diseased' patients, and clinical institutions like Buch—immediately adjacent to Berlin—providing a near endless source of research opportunities, the conditions for neurological research were ripe. In early 1938 Spatz recruited his former collaborator Hallervorden to Berlin-Buch to head the department of neuropathology.

With the advent of the *Aktion* T4 program in January 1940—just two months after Adolf Hitler's signal directive to Brandt and Bouhler—there was now the opportunity to study living clinical presentations along with contemporaneous post-mortem observation. The *Aktion* T4 program of arranged killings allowed for some of the newly established clinical-academic networks to prospectively identify patients with intriguing or significant disorders, arrange a convenient time for their deaths, and then harvest their remains for subsequent study.

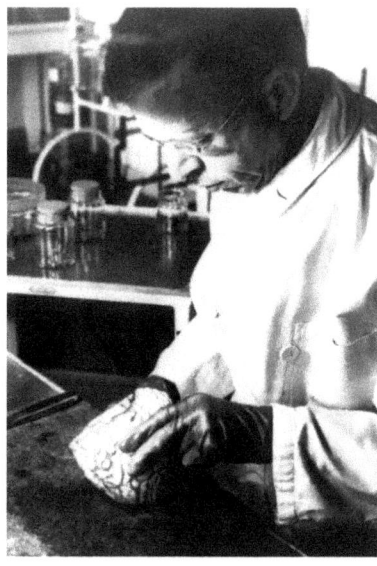

FIGURE 22 Julius Hallervorden

Adolf Hitler's approval of the murder of baby Gerhard Kretschmer in the summer of 1939—and the broader intent behind it—hardened the resolve of the Nazi regime to implement a program of ridding the *Volk* (population) of the weak or hereditarily diseased. Viktor Brack and Herbert Linden, a high-ranking official in the Reich Ministry of the Interior, then established a committee to implement a program of *Kindereuthanasie* called the Reich Committee for Scientific Registration of Serious Hereditary and Constitutional Conditions. This committee included Hans Heinze, Hallervorden's child psychiatrist colleague from the 1920s, now director of Brandenburg-Görden State Hospital.[231]

On 18 August 1939 the Ministry of the Interior issued a confidential circular specifying the kinds of childhood disorders to be considered grounds for 'euthanasia'. The circular directed staff at maternity hospitals, obstetric departments and children's hospitals to report in writing to the appropriate health authorities any children born or under age three with diagnoses of 'idiocy and mongolism', microcephaly, hydrocephalus, congenital malformations of all kinds, and paralysis, including cerebral palsy.[232] Each reporter would receive a small fee for their notification. District medical directors were then required to forward the registration forms to Brack's office in the KdF. A committee of experts was established to review the notifications and decide which of the disabled children were to be admitted to newly established facilities called *Kinderfachtabteilungen* (special children's wards). In the *Kinderfachtabteilungen*, children were to be assessed by paediatricians and child psychiatrists. If the child was assessed unfavourably, they were killed a few weeks after arrival, usually by overdose of Luminal.

Kinderfachtabteilungen were established in at least 30 state hospitals or children's homes throughout Germany, Austria and some of the occupied territories. From 1940 to 1945 more than 5,000 children were murdered in the *Kindereuthanasie* program. In the opinion of the *Reichsgesundheitsführer* (Reich Health Leader) Leonardo Conti, the removal of disabled children enabled families to focus their resources on raising healthy children. The Reich frequently offered financial support to families to help them place the 'defective offspring' in institutions.[233]

The first *Kinderfachtabteilung* in the German Reich was set up in autumn of 1940 in Hans Heinze's Brandenburg-Görden State Hospital. It was around this time that Elvira Hempel was interned there in the hospital's child psychiatry ward, after being sent away from the Brandenburg gas chamber.

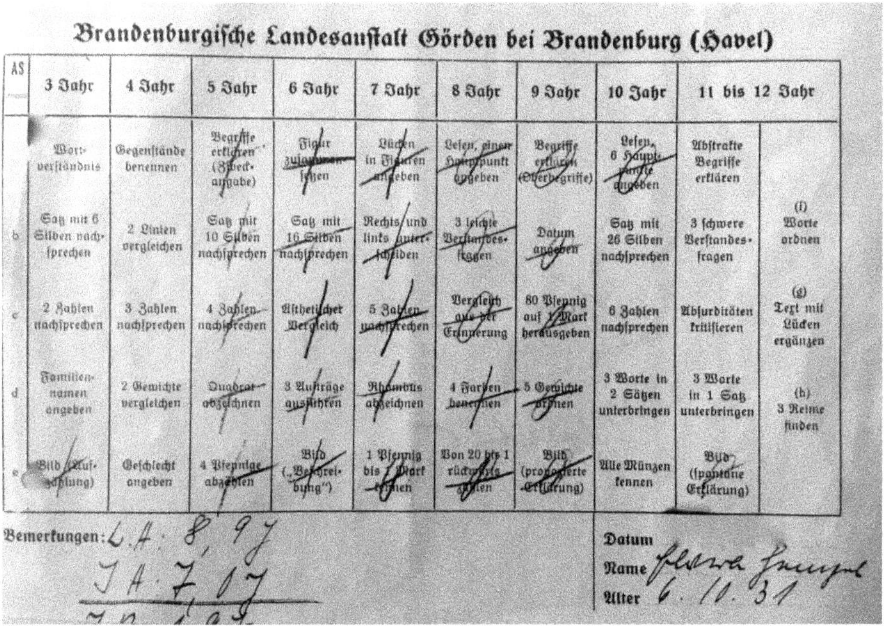

Brandenburgische Landesanstalt Görden bei Brandenburg (Havel)

AS	3 Jahr	4 Jahr	5 Jahr	6 Jahr	7 Jahr	8 Jahr	9 Jahr	10 Jahr	11 bis 12 Jahr	
	Wortverständnis	Gegenstände benennen	Begriffe erklären (Zweckangabe)	Figur zusammensetzen	Lücken in Figuren angeben	Lesen, einen Hauptpunkt angeben	Begriffe erklären (Oberbegriffe)	Lesen, 6 Hauptpunkte angeben	Abstrakte Begriffe erklären	
b	Satz mit 6 Silben nachsprechen	2 Linien vergleichen	Satz mit 10 Silben nachsprechen	Satz mit 16 Silben nachsprechen	Rechts und links unterscheiden	3 leichte Verstandesfragen	Datum angeben	Satz mit 26 Silben nachsprechen	3 schwere Verstandesfragen	(f) Worte ordnen
c	2 Zahlen nachsprechen	3 Zahlen nachsprechen	4 Zahlen nachsprechen	Ästhetischer Vergleich	5 Zahlen nachsprechen	Vergleich aus der Erinnerung	80 Pfennig auf 1 Mark herausgeben	6 Zahlen nachsprechen	Absurditäten kritisieren	(g) Text mit Lücken ergänzen
d	Familiennamen angeben	2 Gewichte vergleichen	Quadrat abzeichnen	3 Aufträge ausführen	Rhombus abzeichnen	4 Farben benennen	5 Gewichte ordnen	3 Worte in 2 Sätzen unterbringen	3 Worte in 1 Satz unterbringen	(h) 3 Reime finden
e	Bild (Aufzählung)	Geschlecht angeben	4 Pfennige abzählen	Bild („Beschreibung")	1 Pfennig bis 1 Mark erkennen	Von 20 bis 1 rückwärts zählen	Bild (prononcierte Erklärung)	Alle Münzen kennen	Bild (spontane Erklärung)	

Bemerkungen: L.A. 8,9 J
I.A. 7,1 J

Datum
Name Elvira Hempel
Alter 6.10.31

FIGURE 23 The developmental assessment conducted on Elvira Manthey

Conducted prior to her transportation to the Brandenburg killing centre, the assessment indicates a cognitive age of 6-7 years old.

Hans Heinze was born in the Saxon town of Elsterberg in 1895. His father, a local businessman and mayor, died when Hans was a child. Heinze served as an orderly in a military hospital during the 1914-1918 war and later studied medicine in Leipzig. After graduating in 1924 he worked in child and adolescent clinics in Leipzig, Berlin and Potsdam. His career flourished at the Potsdam clinic, which he helped turn into a leading institution for child and adolescent psychiatry. Heinze believed that the psychiatric profession had a responsibility to identify intellectually disabled or maladjusted children in the education system, remove them from 'normal' schools and place them in specific institutions.[234] He joined the Nazi party in May 1933.

In 1938 the staff and patients of the Potsdam State Hospital were transferred en masse to the larger Brandenburg-Görden institution, where Heinze was installed as the director. He took up residence in a palatial building in the Brandenburg-Görden complex and appointed Dr Ernst Illing as director of the child psychiatry ward. Illing relocated to the Am Spiegelgrund in Vienna in 1942 and was replaced by Friederike Pusch. In 1939 the Brandenburg-Görden State Hospital had 2,600 beds, of which 1,000 were allocated for child and adolescent patients.[235] The development of different forms of *Heilpädagogik* (curative education) for children with intellectual disabilities was a critical part of therapeutic intervention in German child psychiatry[236] and in this vein Heinze established a *Lebenschule* (school of life) in the Görden institution that sought to assist children with learning difficulties and teach elemental skills in self-care. This enabled the boys to be deployed to provide labour on nearby farms and the girls were sent out to work as domestic servants.

Of the 1,481 children and adolescents admitted to the Görden *Kinderfachtabteilung* between late 1939 and the end of the war, only

436 survived.[237] The children selected for death were gassed in the nearby Brandenburg killing centre. After Brandenburg was decommissioned in late 1940, children were killed on the premises of the *Kinderfachabteilung* by overdose of Luminal. Apart from one documented instance where several epileptic children were 'loaned' to a Luftwaffe research facility in Berlin and subject to low air pressure chamber experiments[238], there is no extant evidence of other children at Görden *Kinderfachtabteilung* being subject to experimentation.

FIGURE 24   Hans Heinze, Director of the Brandenburg-Görden State Hospital, 1938—1945

Elvira Hempel was detained for around three months at Görden in 1941, after which the Brandenburg state authorities refused to bear any further cost for her care and she was returned to Uchtspringe State Hospital. There the murder of children and adults continued. During the 'children's euthanasia' program, a total of 753 children and adolescents were murdered in the Uchtspringe *Kinderfachtabteilung*.[239] Uchtspringe appears to have had a much larger population of children living with intellectual disabilities which may account for its much higher death rate compared to other institutions.²⁴⁰ In addition to the victims of 'children's euthanasia', around 1,800 adult victims of the *Krankenmorde* transited to their deaths through Uchtspringe, which also functioned as a *Zwischenanstalten* (holding centre).²⁴¹ Among these were the elderly women Elvira had observed arriving at the hospital grounds and then later disappearing, including the 52 women transported to the Brandenburg T4 killing centre on the same day as Elvira.

While horrific crimes were being perpetrated at Görden and Uchtspringe, an equally notorious *Kinderfachabteilung* was located at Vienna's *Am Steinhof* psychiatric hospital, where about 800 children were murdered.²⁴² The Steinhof children's psychiatric unit—known as *Am Spiegelgrund*—was first headed by Professor Erwin Jekelius. Jekelius' ruthlessness in selecting children for death and his brutality in applying multiple counts of severe corporal punishment on the children under his care there, led Victor Frankl to describe him as a 'Mephistoclean figure'.²⁴³ Jekelius enjoyed some measure of impunity at the time as it was rumoured he was in a romantic relationship with Adolf Hitler's sister, Paula. However, he was removed from duties at *Am Spiegelgrund* in 1942 and replaced by Ernst Illing, Heinze's former deputy from Görden. Another senior psychiatrist at *Am Spiegelgrund*, Heinrich Gross, was equally relentless in eliminating children with disabilities and later developed an obsessive interest in his collection of brains and tissue samples taken from his victims—a collection that rivalled that of Hallervorden.²⁴⁴ Like Hallervorden, Gross also continued to publish material related to the murder of children under his care for decades after the end of the war.²⁴⁵

The professional relationship between Hallervorden and Heinze flourished during the war. The fact that children, and later adults, with particular case histories could be killed and studied in timed

post-mortem examinations provided a substantial source of research material. One of the functionaries in this process was Heinrich Bunke, Eberl's deputy at Brandenburg, and the large man whom Elvira encountered outside the gas chamber.[246] Bunke had spent a month in Hallervorden's lab in Buch and had been trained in the removal of brains and their preparation for research purposes.[247] On 28 October 1940 the last children's transport from Görden arrived in Brandenburg killing centre.
It comprised at least 58 children, one of whom was Werner Przadka. The transport and post-mortem tissue harvest had been arranged between Heinze at the Görden *Kinderfachabteilung*, Hallervorden at the KWI, and Eberl at Brandenburg. Post-mortem reports and brain tissue samples from 40 of these children were later found in Hallervorden's estate.[248]

Hallervorden and his colleagues were not the only physicians to take advantage of the T4 murders in this way. In January 1941 Carl Schneider, an *Aktion* T4 consultant and professor of psychiatry at Heidelberg, expressed his like-minded opinion that 'the opportunity should not be lost to use it for research on mental diseases... for their therapy and prevention'.[249] Similar to the system established between Hallervorden and Heinze at Görden, Schneider had an agreement with the nearby Eichberg State Hospital to receive the brains of victims. The T4 leadership approved such arrangements. In 1941 the medical director of the T4 program, Paul Nitsche, endorsed these collaborations and specifically recommended that Görden receive all cases of 'congenital idiocy and epilepsy' for study after death.[250] Patients whose brains were to be removed after their murder were marked on the back of their shoulders with a cross drawn with a red crayon just after they had undressed for the last time. Bunke would state in his post-war trial that it would have been irresponsible to have not made use of this research opportunity.[251] Around 2 per cent of the victims of the T4 program were subject to post-mortem examination for research purposes.

Hallervorden alone amassed a collection of more than 700 brains.[252] The number of post-mortem examinations conducted at Hallervorden's lab in Görden increased dramatically during the *Krankenmorde*—from four in 1938, rising to 1,260 post-mortem examinations by 1945.[253] The brains and other tissue samples from T4 victims remained as the 'Hallervorden Collection' at the University of Frankfurt and in the Munich Psychiatric Institute until 1990.[254]

FIGURE 25 A dissection table at the Hadamar killing centre site (present day)
Adjacent to the gas chamber complex, the corpses of victims who were marked with a red pencil were placed on this bench for the removal of brains or other tissue samples of 'research' interest.

Psychiatrist and neurologist Leo Alexander is known primarily for his work advising the prosecution in *United States of America vs Karl Brandt, et al*, conducted at the Palace of Justice in Nuremburg from December 1946 until August 1947. This famous 'Doctors' Trial' of 23 defendants focused primarily on unconsented experiments on prisoners and civilians and included the mass killings under the *Aktion* T4 program.

Alexander was born in Vienna in 1905, the son of an ear, nose and throat surgeon. He was influenced by both his father's medical career and the vibrant multicultural environment of *fin de siècle* Vienna, where Sigmund Freud was often a guest at the Alexander home.[255] He graduated from the University of Vienna Medical School in 1929 and began his psychiatric studies at the University of Frankfurt. He later travelled to China where he became deeply engaged with non-Western approaches to psychiatry and mental illness. Reports of the persecution of Jewish doctors after the rise of the Nazi regime in early 1933 reached Alexander in China and prompted him to seek a new home in the United States, where he gained teaching positions at Harvard and Duke universities. During the Second World War he served as a Major in the US Army in Europe, operating as a specific 'medical investigator'. In 1945 Alexander was appointed as the principal medical advisor to American lawyer Telford Taylor, the chief prosecutor in the series of US-led Nuremberg prosecutions. The Doctors' Trial was one of the subsequent trials at Nuremberg that followed the 1945-46 International Military Tribunal war crimes case against the captured Nazi regime leaders.[256]

Alexander's avuncular appearance belied an incisive intellect and relentless determination to investigate the crimes of his former countrymen under the Nazi regime. Alexander learnt of the activities of Hallervorden through his investigation into the crimes of the medical profession under the Nazi regime. Given Alexander's interests in neurology and neuropathology, it is likely that he knew of Hallervorden prior to leaving Europe in the 1930s.[257]

Alexander travelled to Dillenburg, north of Frankfurt, to meet Hallervorden on 14 June 1945. Hallervorden had moved his lab to Dillenburg a year earlier because of the threat posed by Allied bombing in the Berlin area. Rather than receiving Alexander's visit as an investigation of his alleged criminal activities, Hallervorden thought it was a professional courtesy. He even speculated that Alexander was bringing an offer of financial or material support for his extensive collection of anatomical specimens and as a means of continuing the research program.[258] Alexander's guarded interest in and curiosity toward the extent and detail of Hallervorden's collection is reflected in the report he wrote concerning their meeting, for example: 'Dr. Hallervorden and his co-workers have carried out a great deal of research during the war and he has kept up his high standards as a thorough accurate and ingenious observer'.[259] The tone of the report reflects a level of collegial bonhomie between them that in part explains Hallervorden's misperception of the purpose of the visit. As the conversation with Alexander continued, Hallervorden described his research processes and some of the frustrations, such as inconsistent availability of medical notes and lack of access to other parts of the subject's bodies. Alexander quoted Hallervorden as stating: 'I heard that they were going to do that, and so I went up to them and said, 'Look here now, boys, if you are going to kill these people, at least take the brains out so that the material could be utilised'. They asked me, 'How many can you examine?', and so I told them an unlimited number – the more the better. I gave them the fixatives, jars and boxes, and instructions for removing and fixing the brains and then they came, bringing them in like the delivery van from the furniture company. There was wonderful material among those brains, beautiful mental defectives, mal-formations and early infantile disease. I accepted those brains of course. Where they came from and how they came to me was really none of my business'.[260]

Hallervorden was never prosecuted for his part in the T4 program and his historical legacy has been highly contentious. In 1949 Alexander published an account of Hallervorden's activities during the years of the Nazi regime.[261] From this, word about Hallervorden's wartime conduct spread throughout the international scientific community and by the early 1950s the concerns were ready for further professional examination. With the Fifth International Congress of Neurological Sciences to be held in Lisbon in September 1953, Hallervorden and another controversial German neuroscientist, Georg Schaltenbrand, had been invited as speakers at the meeting. Schaltenbrand's long-term research interest was multiple sclerosis, examining the possibility that the condition had an infectious cause. During the war he had conducted unconsented experiments at his Würzburg laboratory on asylum patients, studying the effects of injecting them with infectious fluids from primates with an artificially induced form of multiple sclerosis.[262]

By all accounts the Lisbon meeting became combative on multiple grounds, including concerns about the organisation of sessions, the quality of many of the presentations, and national bias in the selection of invited speakers.[263] However, while Hallervorden's participation at the Lisbon meeting was divisive, many of his colleagues came to his defence.[264] He was able to attend the meeting and present his work, and subsequently it was Alexander who found himself maligned by the medical profession over his accusations about Hallervorden.[265]

From the end of the war, Hans Heinze continued to work in a clinical capacity at Brandenburg-Görden, then part of the Soviet-controlled East Germany. After refusing an offer by Soviet authorities to take up a professorial position in Crimea, he was tried and convicted of war crimes in March 1946. Heinze spent four years as a prisoner at a special Soviet camp on the site of the former concentration camp at Sachsenhausen. Following release from imprisonment, Heinze took up the directorship of the department of child and adolescent psychiatry in the Wunstorf hospital in Lower Saxony where he spent the remainder of his career. He died in 1983.

Ernst Illing was not as fortunate. While avoiding conviction for murder, he was nonetheless convicted of the 'manslaughter' of many children and executed in 1946. Under a peculiarity of Austrian law, it had been argued that the intellectually disabled children Illing killed could not know they were to die and therefore could not be 'murdered'.[266] Heinrich Gross served only two years for 'manslaughter' under the same anomaly of Austrian law. Erwin Jekelius was captured by Soviet forces in 1945 and convicted of crimes related to euthanasia. He died of bladder cancer in prison in 1952.

Heinrich Bunke returned to medical practice after the war and later trained as a gynaecologist. He was arrested in 1962 and tried for his crimes in the 'Nazi euthanasia' program, but was acquitted. He continued in medical practice until the West German Federal Court sought to retry him in 1970. Bunke was deemed medically unfit to stand trial, yet continued his practice. He was retried in Frankfurt in 1987, convicted of his crimes in *Aktion* T4 and served 18 months' imprisonment. He died in 2001.

FIGURE 26 Leo Alexander assisting the prosecution at the Nuremberg Doctors' trial, 1947—1948
Alexander interviewed Hallervorden in June 1945 in his laboratory at Dillenburg and wrote a detailed report on his wartime activities.

While Leo Alexander is best remembered for his advisory role in the Nuremberg Doctors' Trial, his involvement in the genesis of the ensuing 'Nuremberg code' is his primary legacy in the field of biomedical ethics. The 'Doctors' Trial' revealed the depth of unconsented experimentation on prisoners in camps and on *Ostarbeiter* (foreign labourers) in Germany, experimentation that included imposing and studying the effects of high altitude and freezing, deliberate poisoning, injury or infection, and experimental surgical procedures on human subjects.[267] The 'euthanasia' component of the indictment in the Doctors' trial was overshadowed by these violations in the ethics of medical research, and the post-Nuremburg tradition in biomedical ethics was for a time focused almost exclusively on research ethics.

In May 1946, Alexander proposed six points that might define legitimate and ethical medical research. Following the findings and verdicts of the 'Doctors' Trial' in August 1947, four additional points were added to constitute the ten-point 'Nuremberg Code'[268]:

> Required is the voluntary, well-informed, understanding consent of the human subject in a full legal capacity.
>
> The experiment should aim at positive results for society that cannot be procured in some other way.
>
> It should be based on previous knowledge (like, an expectation derived from animal experiments) that justifies the experiment.
>
> The experiment should be set up in a way that avoids unnecessary physical and mental suffering and injuries.
>
> It should not be conducted when there is any reason to believe that it implies a risk of death or disabling injury.
>
> The risks of the experiment should be in proportion to (that is, not exceed) the expected humanitarian benefits.
>
> Preparations and facilities must be provided that adequately protect the subjects against the experiment's risks.
>
> The staff who conduct or take part in the experiment must be fully trained and scientifically qualified.
>
> The human subjects must be free to immediately quit the experiment at any point when they feel physically or mentally unable to go on.

Likewise, the medical staff must stop the experiment at any point when they observe that continuation would be dangerous.

These principles were only reified in 1964 in the 'Declaration of Helsinki', when the World Medical Association absorbed most (but not all[269]) of them into its statement of ethical principles for research involving human subjects. However, the World Psychiatric Association (WPA) would not engage with a universal code of ethics until concerns around the misuse of psychiatric diagnosis and treatment in persecuting political opponents or dissidents in the Soviet Union were resolved, leading eventually to the 'Declaration of Hawaii' in 1977.

Following the WMA's 1964 Helsinki Declaration on research ethics, and a growing realisation of the problems raised by the crimes of those like Hallervorden and Gross—who continued to occupy academic positions in Germany and Austria—debate emerged about the ethics of publishing academic papers based on illegal and unethical research activities in the Nazi period.[270] Around 30 neurological syndromes were named after physicians who were in some way linked to the Nazi regime.[271] Just as pantothenate kinase-associated neurodegeneration is no longer referred to as 'Hallervorden-Spatz disease', there was increasing pressure to abandon other such 'eponymous' diagnostic labels.[272] Hans Asperger, whose name is well known in association with the spectrum of mild autism disorders in children and adults, was head of the Department for Special Education / Orthopedagogy of the University Paediatric Clinic in Vienna from 1932 until 1940 when he served as a medical officer in Nazi occupied Yugoslavia. As a 'special education consultant', he had provided assessments of children's 'educability' as part of a commission. Asperger had assessed as 'uneducable' 35 children at the Gugging paediatric institute. All but six of these children were sent to *Am Spiegelgrund* in Vienna, where none survived.²⁷³ As historian Herwig Czech noted: 'Asperger managed to accommodate himself to the Nazi regime and was rewarded for his affirmations of loyalty with career opportunities. He joined several organizations affiliated with the NSDAP (although not the Nazi party itself), publicly legitimized race hygiene policies including forced sterilizations and, on several occasions, actively cooperated with the child 'euthanasia' program'.²⁷⁴

In the Fifth Edition of the American Psychiatric Associations Diagnostic and Statistical Manual of Mental Disorders (DSM 5 published in 2013), the former 'Asperger's disorder' has been re-named 'autism spectrum disorder'.²⁷⁵

The Nuremberg Code did not engage with the *Krankenmorde* and it was not until 1996 that the WPA's 'Declaration of Madrid' extended the scope of universal proclamations of psychiatric ethics and addressed specifically the issue of 'euthanasia'.²⁷⁶ The resulting 'Declaration of Madrid' (1996) proclaimed that 'a physician's duty, first and foremost, is the promotion of health, the reduction of suffering, and the protection of life. The psychiatrist, among whose patients are some who are severely incapacitated and incompetent to reach an informed decision, should be particularly careful of actions that could lead to the death of those who cannot protect themselves because of their disability. The psychiatrist should be aware that the views of a patient may be distorted by mental illness such as depression. In such situations, the psychiatrist's role is to treat the illness'.²⁷⁷

In 1948, the Kaiser Wilhelm Institute was dissolved and reformed as the *Max Planck Gesellschaft* (Max Planck Society, the 'MPG') and the KWIHF became the MPG Institute for Brain Research. Hallervorden continued there as director until his retirement in 1955, whilst Spatz stayed in his research role until retiring in 1959. Hallervorden's acolyte, Wilhelm Krücke, was director of the MPG's Institute for Brain Research until 1981. Following Hallervorden's retirement, the institute refocused its research efforts on non-clinical neuroscience with two new departments established: neuroanatomy (headed by Heinz Wässle) and neurophysiology (headed by Wolf Singer). Wässle and Singer's directorships were to be troubled by the long shadow of Hallervorden.

Both Wässle and Singer told filmmaker Jasmine Wingfield in 2016 that after Krücke's retirement they became aware of the presence of numerous brain sections and tissue samples of unknown provenance in their institute.[278] In 1984, Singer had granted informal access to the institute to Götz Aly, then a left-wing political activist with the *Rote Hilfe* ('Red Help') movement. During his academic studies, Aly had encountered the work of neurologist and medical historian Jürgen Pfeiffer in this area and had become interested in Hallervorden and his post-war activities in the MPG. By comparing the dates of death on some of the specimen files, Aly identified them as coming from 33 of the 38 children from Görden killed at Brandenburg killing centre on 28 October 1940, presumably at the request of Heinze and Hallervorden.[279] Aly's interest in this area was not welcomed by many senior scientists at the MPG as the institution was already under siege by anti-vivisectionist groups. Many MPG staff agitated for legal action against him, although Singer recognised this would only make a bad situation worse. Moritz Helmstädter, one of the subsequent directors of the MPG at the time of Wingfield's interviews in 2016, argued that his colleagues seemed more concerned that the institution's research output would be rejected by American scientific publications and its researchers made unwelcome at international scientific meetings, than presenting any sense of guilt about Hallervorden's crimes or the MPG continuing to benefit from his war-time activities.[280]

The first inclination of Wässle and Singer was to lock away the specimens and hope the matter would fade from view. However, the scandal grew in the media and both the Israeli government and the political left in the West German parliament agitated for a response. In 1989, then German Chancellor Helmut Kohl directed the MPG and other scientific institutes with similar specimens to provide them with proper burial. The University of Tübingen's department of neuropathology issued a public apology on 11 January 1989 and then convened a full inquiry into its activities during the Nazi period.[281] The review identified that Hallervorden had collected tissue specimens from 1,540 victims of the Nazi regime and Hugo Spatz—his former collaborator from the 1920s—had amassed a collection of tissue from more than 1,400 people. The provenance of the tissue samples was identified as coming from more than 630 victims of *Aktion* T4, in addition to tissue from executed prisoners, concentration camp inmates, Jewish people who had died in ghettos, German military personnel and a small number of allied prisoners of war.[282]

In May 1990 more than 84,000 slides from 1,540 victims from the 'Hallervorden Collection' were buried in aluminium canisters in Munich as part of a multiple denominational ceremony—Protestant and Catholic clergy were present at the ceremony as was a rabbi, even though there was no knowledge of whether any of the victims were Jewish. Selected public dignitaries were invited to the event but not relatives of the victims, nor representatives from victim groups, disability or mental health NGOs, nor interested members of the public. The victims were commemorated as victims of the Nazis, not victims of the scientists who profited from the Nazi regime. British historian Paul Weindling was intensely critical of the manner of burial of the tissue samples, which he saw as a panicked response. Weindling argued that the lack of any attempt to identity the children whose brains were acquired after their murder was little more than a continuation of the process of dehumanisation at the time of their deaths.[283]

In 1997 the then president of the MPG, Hubert Markl, appointed a committee of independent historians to investigate the history of the KWIHF during the Nazi period; this investigation ran from 1998 to 2004. Despite the MPG's refusal to comply with Weindling's recommendation of a full-scale analysis of all the MPG's holdings in its psychiatric research centre, the multi-volume report presented irrefutable evidence of widespread collaboration with the Nazi regime and career opportunism by Hallervorden and other scientists.[284] Markl then presented a formal apology for the crimes perpetrated by KWIHF scientists[285], although this was in itself a controversial act as 'there was unmistakable disagreement over the correct relationship between apology, forgiveness, coming to terms with the past and remembrance'.[286] The multiple components needed for the MPG to engage with its Nazi past could not be simply bundled together in one act of contrition.

The MPG now hosts a permanent art exhibit remembering the crimes committed by the KWIHF under the Nazi regime. In October 2015, the MPG held a commemorative event 'Provenance and Personal Identity: Problems of Brain Specimens and Tissues from the Era of National Socialism' in honour of the anniversary of the deaths of the 38 children whose brains were identified in the Hallervorden collection by Götz Aly. The MPG invited its harshest critic, Paul Weindling, as a keynote speaker at the event.[287]

Prior to accepting the invitation, Weindling insisted that all the 38 murdered children be named, including Werner Przadka.

In May 2017, a group of Holocaust scholars met in the Israeli city of Akko. At this meeting, Paul Weindling reported on recent developments in the MPG's engagement with its past. Weindling advised that the MPG had commissioned another international panel of historians to progress the previous investigation that had ended in 2004. He indicated that the MPG was demonstrating a considerable degree of institutional inertia, predominately due to legal advice. He told the Akko meeting that the newspaper *Suddeutsche Zeitung* (South German Newspaper, 25 April 2017) had reported that three cases of microscope slides containing brain material from victims of the *Krankenmorde* had gone missing (presumed stolen) and that it was also probable that the MPG's Berlin laboratory still held tissue samples of victims.[288]

CHAPTER 6
'CURE OR DESTROY'

Throughout the 20 months of *Aktion* T4, specially selected doctors from across Germany and Austria travelled to Berlin to an elaborate Italianate villa on the *Tiergartenstraße*. This important transport avenue formed the southern boundary of the extensive *Tiergarten*, the vast parkland in central Berlin. At the northern edge of the park lay the Reichstag, the seat of government power now known as the German Bundestag. Barely one kilometre away, at *Tiergartenstraße* 4, the doctors arrived to make their clandestine contribution to the health of the Nazi nation.

The committee of medical assessors that gathered in the *Tiergartenstraße* 4 office to adjudicate over the future of Lisa and Elvira Hempel and hundreds of thousands of other 'genetically inferior' Germans was chaired by Werner Heyde, an academic psychiatrist from Würzburg in Bavaria and member of the SS. Heyde's deputy and the physician responsible for coordinating between the T4 office and the numerous institutions where potential victims were located was Hermann Paul Nitsche. After December 1941, Nitsche became the medical director of the T4 Medical Committee. Nitsche's trajectory as physician and psychiatrist through the malevolent medical practices of the Nazi euthanasia program personifies the evolution of the psychiatric profession's values in the first half of the twentieth century, many of which remain influential—and controversial—in the present day.[289]

Sigmund Freud and Karl Jung are often considered the dominant figures of the history of psychiatry in the early twentieth century. Lesser known is the work of Emil Kraepelin. Freud and Kraepelin represent antitheses in psychiatry and embody a tension in the discipline that continues to trouble it. From the *fin de siecle*, the German psychiatric profession had divided along the lines of those who viewed mental illness as either a biological phenomenon (*Somatiker*) or a disease of the mind (*Psychiker*). With the rise of the Nazi regime in Germany and the enactment of a broad policy of purging all the professions of Jewish influence and bringing them into line with the interests of the state, Freud's ideas were rejected and demonised as the foundation of the 'Jewish Science' of psychoanalysis.[290] This void enabled Kraepelin's legacy to profoundly influence German, and ultimately Western, psychiatry.[291]

Kraepelin was long interested in the classification of psychiatric disorders and in establishing a paradigm of biological psychiatry. Biological psychiatry held that mental illness was the result of pathological processes affecting the brain. This drew Kraepelin to social Darwinism, eugenics and racial hygiene, through which he formulated a view that modern humanity and its supportive medical, welfare and educational policies was undermining true natural selection and allowing inferior genetic traits to flourish. After Kraepelin died in 1926, his biological psychiatry project continued through the work of his protégé, the radical eugenicist psychiatrist Ernst Rüdin. Rüdin helped formulate the 1933 Hereditary Health Laws and later advocated for the killing of genetically inferior humans.[292]

Psychiatry in Germany suffered a poor reputation in the medical and general community before the 1914-1918 war and an even worse one following it. Psychiatrists seemed unable to provide any real relief for the suffering of returned soldiers, deeply traumatised by trench warfare. Rather than healing physicians, psychiatrists were considered as little more than custodians of large and costly asylums. In Germany these asylums consisted of *Heilanstalten* (sanatoria) and *Pflegeanstalten* (nursing homes or care institutions) which had become depositories for a vast number of patients suffering severe and disabling mental illness and degenerative neurological diseases. The mix of patients in asylums made the 'curability' of this population much more difficult than that of patients in other fields of medicine. The rapid pace of modernisation and profound changes within the social and economic order in Germany after unification in 1871 forced many people into cities in search of work. These unsettling fluctuations in German society caused a spike in psychological distress and mental illness.[293] From the late nineteenth century until the German defeat in 1918, the number of inmates in various psychiatric institutions in Germany increased five-fold. However, the exponential growth of asylum populations in Germany was not met by increased government resourcing. During the 1914-1918 war, the privations arising from the British naval blockade of foreign imports of food and other necessities into Germany contributed to the death of more than 70,000 German psychiatric patients—around one third of the total asylum population.[294]

In the early part of the twentieth century, being committed to a psychiatric institution in Germany was primarily at the request and discretion of the patient's family, as was the case in many other countries. There was little in the way of legal or clinical challenge to such admission into an asylum and many such commitments were sought on the grounds of avoiding social embarrassment to a family. However, during the Weimar Republic years (1919-1933), the comparatively liberal society and culture that flourished briefly created a vigorous anti-psychiatry sentiment in the German community, comparable to those seen in the social liberation movements in the United States and Europe in the 1960s and 1970s. An assertive patient's rights movement combined with social and political demands for more legal oversight of the process of involuntary commitment generated unprecedented challenges to the profession.[295] There followed an ultimately failed experiment in de-institutionalisation, with attempts to deliver psychiatric care in the community.[296] As was the case in other western countries, the asylum remained the locus of psychiatric power and as the new social psychiatry experiment floundered, so the setting of psychiatric care gradually returned to the asylum model in the latter years of the 1920s.[297]

The apparent 'incurability' of many patients remained a challenge in the asylum system. With the rapidly deteriorating financial and political situation in the Weimar Republic, a form of moral panic found expression within the broader community generating an enthusiasm for eugenics.[298] Mirroring similar social currents in the United States and other parts of Europe, the German government and public increasingly despaired of the financial burden of 'useless eaters' in costly asylums and the perceived unrestrained reproduction of a potential genetic underclass. This notion of a growing, interbreeding subclass, corrupting morality and dragging the country down, would become a persuasive metaphor. Similar sentiments were influencing support for policies of compulsory sterilisation of the 'genetically inferior' in other countries.[299]

Not all responses to the burgeoning asylum population were as condemning or restrictive. A 'reform movement' in German psychiatry[300] sought to capitalise on Kraepelin's paradigm of biological psychiatry through the introduction of a range of physical treatment in asylums. These included malaria therapy[301], aversion therapies[302], insulin coma therapy and early applications of convulsive therapy using cardiazol or rudimentary electroconvulsive therapy (ECT) equipment.[303] Some German psychiatrists, such as Anton von Braunmühl at Eglfing-Haar, collaborated closely on the development of commercially viable ECT machines with the medical device company Siemens-Reiniger-Werke[304], after which the Siemens 'Konvulsator' appeared in German psychiatric clinics in 1940. The Vienna-based medical device company F Reiner and Co later developed the Elkra I and II ECT devices for use in Austrian psychiatric hospitals.[305]

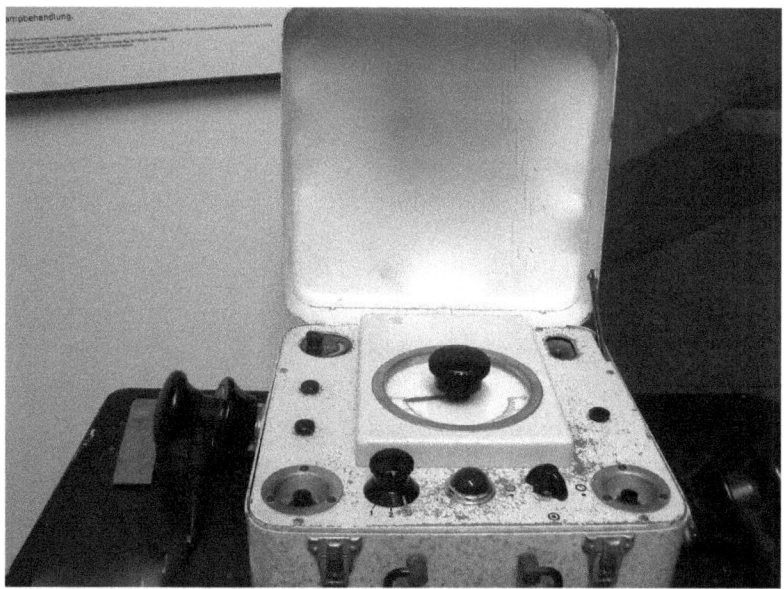

FIGURE 27 A prototype Electroconvulsive Therapy (ECT) machine
This ECT machine was used in German psychiatric hospitals in the 1940s.

Prior to the introduction of biological treatments, patients in asylums were subject to prolonged bathing, physical restraint or rapid movements in special swing contraptions or spinning chairs. Alongside the introduction of these new treatments, a Bavarian psychiatrist, Gustav Kolb, sought to revisit the social psychiatry experiment and established a clinical program known as *Offene Fürsorge* (open care) during the early years of the Weimar republic. Kolb was as determined as the asylum reform psychiatrists to change psychiatric care and his model of outpatient treatment within the broader community became a defining characteristic of public mental health care before the Nazi regime.[306] Hermann Simon, another asylum director, sought to bring *Arbeitstherapie* (work therapy) into the asylum system.[307] *Arbeitstherapie* usually involved placing male patients in agricultural and physical labouring jobs, and female patients in kitchens, laundries or process work. In some asylums, *Arbeitstherapie* provided a valuable revenue stream. At the Christophsbad Asylum in the town of Göppingen near Stuttgart, the bottling of high quality mineral water by patients provided a lucrative source of income for the institution.[308] By 1933, up to 80 per cent of asylum patients in Germany were active in some form of productive labour.[309]

Arbeitstherapie resembled the 'moral treatments' for asylum patients first introduced in the nineteenth century by the Tuke family at the York Retreat in England.[310] The Tuke's approach to the care of 'madness' included providing asylum patients with social roles, moral instruction and structured recreation time. These ideas and applications—drawn from Quaker ideals—spread to the United States and Europe and their clinical outcomes were particularly impressive. Between 35 and 80 per cent of patients in the York Retreat were discharged home 'substantially improved' or even 'cured'.[311] However, the German asylum reform movement, particularly the central place of *Arbeitstherapie* in the process, had created what was to become a deadly paradox. Rather than enhancing the care of all asylum patients, *Arbeitstherapie* demonstrated that while many patients were capable of being productive, another group of patients were unable to work at the same level and there appeared to be little that could be done to improve their situation.[312]

From this dichotomy came the recognition of those patients who could be 'cured' and those who could not. As social unrest and moral panic spread through German society in the aftermath of the First World War and the Great Depression, the psychiatric profession and general community increasingly believed that this latter group should be excluded from their community. In many ways this sentiment was in keeping with the latest international social, scientific and legal policy developments in the public health area known as 'eugenics' and 'mental hygiene', but in the setting of a new and pernicious public policy under the incoming Nazi regime, exclusion also meant destruction.[313]

―――

Hermann Paul Nitsche was born in 1876 in Colditz, near Leipzig in Eastern Germany. His father, Hermann senior, was a psychiatrist who worked at the Sonnenstein asylum in Pirna on the outskirts of Dresden, the future site of one of the six *Aktion* T4 killing centres. After attending elementary school in Pirna, the younger Nitsche completed his schooling in Dresden in 1896 then studied medicine in Leipzig, Berlin and Göttingen, graduating in 1901. Nitsche was drawn ultimately to his father's specialty of psychiatry and his final year dissertation was on the subject of cognitive impairment in organic brain disease.[314] He commenced training in psychiatry in Frankfurt in 1902 and in 1904 accepted an offer to work in Heidelberg with Kraepelin and Rüdin. During this period of professional development in Heidelberg, Nitsche fell under the influence of the founder of the German 'racial hygiene' movement, Alfred Plötz, and was inculcated in the tenets of biological psychiatry.[315] Nitsche's reputation as a clinician and administrator grew and in 1907 he was offered a senior position in the Municipal Mental Hospital in Dresden.[316] While in Dresden, Nitsche enacted his version of asylum reform through the introduction to the institution of physical treatments such as malaria therapy and insulin coma therapy. In 1913, in recognition of his accomplishments in asylum reform, Nitsche was appointed deputy director at his father's former workplace, the institution at Pirna-Sonnenstein.

FIGURE 28 Herman Paul Nitsche

The former castle at Sonnenstein had been converted into an asylum in 1811 and functioned as an institution for 'curable' patients, a *Heilanstalt*. Sonnenstein had enjoyed a good reputation, particularly for its reforms, although not all accounts of the institution were favourable. The asylum figures prominently in the famous memoir of Daniel Paul Schreber.[317] Schreber was a well-respected jurist and Chairman of the state court of Saxony.

Part 1 - The Crime

In 1884, aged 42, Schreber experienced an episode of paranoid psychosis and was admitted to a psychiatric hospital in Leipzig. He subsequently suffered further episodes and later spent eight years in the Pirna-Sonnenstein asylum. In his memoir, Schreber recounted being brutalised by *Pfleger* (male nursing attendants) at Pirna-Sonnenstein, including having his head held submerged in a full bath, being woken by having his beard pulled violently, and being thrown against his bed. Schreber wrote: 'Sometimes I opposed such indignities with actual resistance, particularly when one wanted to remove the wash-basin for the night from my bedroom, which was locked from the outside; or when one tried to move me from my own bed-room to sleep in the cells fitted out for raving madmen. Later on I desisted from all opposition because it led to senseless scenes of violence; I kept silent and suffered'.[318] Schreber later argued successfully in the Superior Court in Dresden against his further commitment and was released from Pirna-Sonnenstein in 1902. However Schreber's mental health deteriorated again in 1907 and he was re-hospitalised in Leipzig until his death in 1911.[319]

The timing of Nitsche's appointment at Pirna-Sonnenstein also proved significant for other reasons. The fate of asylum patients at the institution during the privations of the 1914-1918 war was particularly grim—the death rate at Pirna-Sonnenstein rose alarmingly from around 5 per cent per annum in 1913 to 32 per cent by 1917.[320] This experience appeared to kindle in Nitsche something of a Malthusian view of the situation in Pirna-Sonnenstein. He sought to improve the situation there by introducing better nutrition in the diets of 'treatable' patients whose illnesses were 'curable' and who could work. In contrast, those patients considered too disabled to be economically productive received severely restricted rations. To provide adequate resources for the 'curable' in times of hardship, Nitsche reported to the Saxon authorities inflated numbers of patients at Pirna-Sonnenstein so as to secure more food, heating oil and other necessities.³²¹ In 1928 Nitsche was promoted to director of the Pirna-Sonnenstein asylum and appointed to a professorship by the State of Saxony the following year.³²²

FIGURE 29 The former Heilanstalt at Pirna-Sonnenstein (present day)
The building now houses the documentation centre and memorial.

With the subsequent rise of the Nazi government, all social institutions and professions were compelled to adopt the aims and values of the regime.³²³ The *Gleichschaltung* (alignment) of the medical profession was based upon notions of public health and hygiene and the welfare of the *Volk* (the 'Aryan' population).³²⁴ Robert J Lifton, whose book *The Nazi Doctors* is a foundation text of the field, described the *Gleichschaltung* as 'a euphemism for eliminating all possible opposition, whether by exclusion, threat, or violence'.³²⁵ Lifton observed that many in the medical profession in Germany were enthused by the advent of National Socialism, particularly as it placed biology and the health of the *Volk* as the highest priority and thus situated the profession in a profoundly influential position. By contrast, racial, biological or political undesirables, such as Jewish doctors, were subject to exclusion from the new German social order in a process referred to *Ausschaltung*.[326] In addition to the prominent role to be played by the German medical profession in the creation of a racial state, the Nazi regime gifted German doctors many of their long-held demands. In 1935 the medical profession was given the right of professional self-governance, an aspiration dating from the 1870s. The German medical profession subsequently established a new peak body, the *Reichsärztekammer* (Reich Physicians' Chamber).[327] This and other concessions in funding, endorsement of medical research institutions and increased political power, made the Nazi regime very attractive to German doctors of whom approximately 60 per cent joined either the Nazi party or its associated bodies—by far the highest proportion of any professional group in Germany.[328] Around 70 per cent of psychiatrists joined Nazi affiliated organisations.[329] Many psychiatrists harboured views about their discipline that were broadly consistent with the aspirations of National Socialism and had joined the Nazi party prior to Hitler securing power in 1933. Nitsche became a member in that year.

During the 1930s, Nitsche opted to serve as a consultant to the Saxony-based hereditary health courts and ensured that as many patients as possible at Pirna-Sonnenstein were sterilised. He was also an office bearer for the German Association for Psychiatry and was instrumental in the process of uniting it with the German Association of Neurology in 1935. After the invasion of Poland in September 1939, Nitsche served as an advisory psychiatrist to the German 10th Army Group operating in Silesia in the country's south west.[330]

Just after the outbreak of the war, the Saxon government—under direction from Berlin—resolved that the Pirna-Sonnenstein asylum should be closed and the facilities used as a military hospital. Patients were dispersed to other state institutions, although a section of the hospital continued as a smaller scale psychiatric hospital under the name 'Mariaheim'. The complex later housed ethnic German refugees from Bessarabia, present day Moldova and Ukraine. This pattern of closing psychiatric hospitals to make way for military hospitals and convalescent accommodation continued throughout the war. From 1943, psychiatric hospitals, by virtue of their remote location and lack of damage from Anglo-American bombing, were sporadically cleared and their occupants killed in a process often referred to as *Aktion Brandt*, named after Dr Karl Brandt.[331]

With the commencement of *Aktion* T4 in 1940, Gustav Kaufmann, chief of the T4 Inspectors Office, appropriated buildings 1-3 at the front of the Pirna-Sonnenstein complex. Drawing on the experience of constructing the Brandenburg killing centre, Kaufmann supervised the building's conversion into a killing centre throughout the first part of 1940. Once operational, Pirna-Sonnenstein was code-named 'Centre D'. Horst Schumann, the former director of Grafeneck (code-named Centre A), assumed the role of chief physician at Pirna-Sonnenstein and the killing operations began in June 1940.[332] The Pirna-Sonnenstein killing centre continued operating as a site of *Aktion* T4 and then *Aktion* 14f13 until September 1942, utilising around 100 dedicated T4 staff during this time.[333] In all, 13,720 victims of *Aktion* T4 were murdered at Pirna-Sonnenstein[334] and a further 1,031 Jewish, Polish, Czech and German concentration camp prisoners were killed as part of *Aktion* 14f13.[335]

Despite the clandestine activities and complicity of numerous medical professionals in hiding the true purpose of such killing facilities, knowledge of the murders seemed commonplace in the community. Victor Klemperer's diary noted on 21 May 1941 that: 'Sonnenstein has long ceased to be a regional mental asylum. The SS is in charge. They have built a special crematorium. Those who are not wanted are taken up in a kind of police van. People here call it "the whispering coach". Afterwards the relatives receive the urn. Recently one family here received two urns at once'.[336]

Nitsche was not directly involved in the day-to-day killing operations at Pirna-Sonnenstein. However, at the state research centre at Leipzig-Dösen, he conducted experiments with Luminal and demonstrated that a dose of 300mg of the drug administered to victims three times daily was enough to facilitate death in a few days.[337] In total, around 60 patients were killed at Leipzig-Dösen in Nitsche's Luminal experiments.[338] The results of these experiments were later adopted in the decentralised euthanasia phase of the *Krankenmorde* and became known as the *Luminalschema*. Nitsche had been one of those present at the initial gassing demonstration at Brandenburg an der Havel in January 1940, and he sought to compare this killing method with that of his Luminal experiments.[339]

Nitsche and others in the T4 leadership explored further killing methods including electrocution. At Gugging hospital in Klosterneuburg on the outskirts of Vienna, a general practitioner, Emile Gelny, developed a means of killing patients by electrocution by using a modified Elkra II ECT machine. Electrocution presented a cheaper method of mass killing than the increasingly harder to source barbiturates. It also provided a useful deception, as the killing could be disguised as a therapeutic measure. There is evidence of substantive discussions between Brandt, Nitsche and Gelny about the widespread adoption of ECT machines as a means of killing patients in psychiatric hospitals. It is probable that Nitsche was also in attendance at a demonstration of the ECT machine as a killing device at Gugging Hospital in the summer of 1944.

Gelny is estimated to have killed around 400 patients by electrocution at Gugging. After the Red Army occupied the area around Klosterneuburg in April 1945, Gelny moved his ECT killing device to another hospital, Mauer-Öhling in between Vienna and Linz, and murdered another 140 people. Gelny fled to Syria at the end of the war and later practiced medicine in Iraq. He died in Baghdad in 1961.[340]

———

In May 1940, Nitsche joined Werner Heyde at the *Tiergartenstraße* 4 office in Berlin as a member of the T4 Medical Office.[341] Nitsche succeeded Heyde as leader of the T4 Medical Committee in December 1941, just after *Aktion* T4 had ceased and the killing process was dispersed to hospitals.[342]

While Nitsche's main priority in this period was the destruction of 'life unworthy of life', he and psychiatric colleagues Ernst Rüdin, Max de Crinis, Carl Schneider and Hans Heinze remained determinedly engaged in the development of the psychiatric profession in Germany, including teaching and research and publishing their work in reputable journals such as *Der Nervenarzt*.[343] Like Julius Hallervorden—whose crimes were discussed in the previous chapter—Nitsche saw merit in capitalising on the research opportunities provided by the availability of a large amount of brain tissue from the victims of the *Aktion* T4 program.[344]

———

Part 1 - The Crime

The convoluted lives of the Mall brothers—Georg, a young man with a mental illness, and Gerhart, training in medicine and psychiatry—provides another disturbing insight into the relationship between the psychiatric profession and its untreatable patients in the Nazi period.[345] Gerhart Mall was born in February 1909 in Codacal in the then British colony of India. Georg was born four years later. The boys' parents were then working as missionaries for a Swiss organisation but as war between Britain and Germany appeared imminent, they returned to Germany in 1913. The Malls lived briefly in Baden before settling in Gmünd, 50 kilometres east of Stuttgart in Southern Germany.

Whilst Gerhart commenced medical studies at the University of Tübingen in the summer of 1931, his brother Georg's life followed a more unorthodox path, drifting from one undertaking to the next until he enrolled in protestant theology, also in Tübingen. Georg soon struggled with his theological studies and became severely depressed and suicidal. He wrote in November 1932 that he wished to discard life like a 'half-eaten apple'. He was admitted to the psychiatric clinic in Tübingen in December 1932 after he had harmed himself in a suicide attempt. At the time of the admission he was malnourished, anxious and paranoid. Georg's family took him home after a few days in hospital, however he remained unwell and was re-hospitalised in Tübingen the following February. After poor clinical progress, his brother Gerhart, still a medical student, requested that Georg be transferred to the Christophsbad clinic in Göppingen, nearer to the family home.

The admitting doctor at Christophsbad noted of Georg on April 22 1934: 'His illness became apparent when he began to suffer from mood swings and started to blame his parents for his bad character. His facial expression became disturbed and he was absent-minded. His mood changed from agitated to happy. All of a sudden he wasn't sure whether he wanted to continue studying theology, went to lectures on medicine instead, didn't know what he wanted and became aggressive. In the end he refused to go to bed at home and watched other family members while they slept. When he realised that they were still breathing, he relaxed. The patient locked himself into a room with a kitchen knife and said he had to watch out, something would happen later. The patient is described as completely stuporous and negativistic. He stands in the room in a rigid stance and is unable to answer. When the nurse came to collect him he actively resisted and cried for help'.[346] Georg's medical record states that he was diagnosed with a schizophrenic illness. The Christophsbad psychiatrist also seemed concerned about Gerhart Mall's demeanour, later writing in the file: 'The patient's brother gives a clearly eccentric and almost blocked impression. He doesn't seem to be able to give any information about his siblings, he can't remember anything. Claims to be a medical student in his 9th semester...He was unable to understand how such an intelligent person could become so ill'.

FIGURE 30 Georg Mall

In July 1933 Gerhart had written to Georg's psychiatrist asking him to explain what Georg's diagnosis of schizophrenia meant and specifically whether it was hereditary. Gerhart's increasing involvement in Georg's situation is evident throughout the Christophsbad medical file. There are accounts of complaints against staff, questioning the diagnosis and its features and, on one occasion, a demand to take Georg to another facility, a move not opposed by Dr Karl John, Christophsbad's medical director.

Gerhart Mall took his final medical exams in January 1935 and commenced an eight-month internship at the psychiatric clinic at the University of Tübingen. Gerhart's superior at Tübingen, Professor Hermann Hoffmann, was an advocate of insulin coma therapy for schizophrenia. Hoffmann, like many psychiatrists, viewed schizophrenia as an *Erbkrankheit* (hereditary disease) and endorsed the sterilisation of these patients.

While Gerhart's life and career flourished, Georg's mental health deteriorated catastrophically. His medical file consistently reports his disorganisation, neglect of self-care, incoherent mumbling and occasional outbursts of unprovoked violence. Gerhart insisted that the hospital administer insulin coma therapy which they initially refused. In mid-January 1937, Gerhart took Georg to the Tübingen clinic where he worked; however, after a series of insulin shocks, Georg showed no signs of improvement and was returned to Christophsbad in March 1937. He then received a series of convulsive treatments using Pentylenetetrazol[347] with so little benefit that his treating psychiatrist wrote in August 1937: 'Patient cannot be saved mentally, no contact to the outside world'. Georg's severe hebephrenic schizophrenia continued with frequent outbursts of violence and profound neglect of his self-care.[348] Georg then underwent what would seem to be a last ditch, but ultimately failed, attempt at treatment with malaria therapy in

June 1938. By June 1940 his behaviour was reported as extremely difficult and his file states that he was transferred to the hospital at Weißenau on 26 June 1940, 'by order of the Ministry'. Weißennau functioned as a *Zwischenanstalt* for the Grafeneck killing centre from which 691 victims were sent to their deaths during the *Aktion* T4 period.[349] On 5 December 1940, Georg Mall was placed on a grey *Gekrat* bus and taken to Grafeneck where he too was murdered—one of 10,654 victims of Grafeneck's gas chamber.[350]

In 1997 journalist Hans-Joachim Lang located two revealing letters written about the death of Georg Mall.[351] One letter from Gerhart Mall to his colleagues, dated 5 October 1940, requests that Georg be 'channelled towards euthanasia' in the Weißennau hospital (presumably by overdose) and not by poison gas at Grafeneck. In response, a letter from the Interior Minister of Württemberg advises Gerhart that he was mistaken in his understanding of 'the usual path' (meaning death in the gas chamber) which the Minister claimed had been 'subject to phantastical rumours' and was a 'no less decent' means of dying than other methods.

Gerhart completed his higher specialist degree in March 1942 and was later appointed a lecturer at the University of Marburg, where he remained until the end of the war. The US occupation forces briefly suspended him from practice, however it seems that after he received his denazification certificate he was reinstated to his position. He returned to Tübingen in 1946 and was made professor of psychiatry in 1949. A few years later he became director of the Palatine *Landesklinik* in Landeck. In 1967 he published a pictorial book of images of patients with hereditary psychiatric disorders.[352] Many of the images in the book are redolent of those used in the

1930s propaganda films such as the 1937 *Opfer der Vergangenheit: Die Sünde wider Blut und Rasse* (The victims of the past, the sins of blood and race).[353] Gerhart Mall died in 1983.

Gerhart is easily cast in the role of zealous eugenic psychiatrist—procuring the 'mercy death' of a genetic inferior, while trapped in his own fears of heritable insanity. He evolved from a close and concerned relative to a dispassionate *Aktion* T4 co-conspirator, at least in the eyes of Hans-Joachim Lang. In many ways his narrative travels a similar arc to that of Paul Nitsche—from the optimism of biological psychiatry as a treatment for the curable patient, to a murderous nihilism towards the incurable patient.

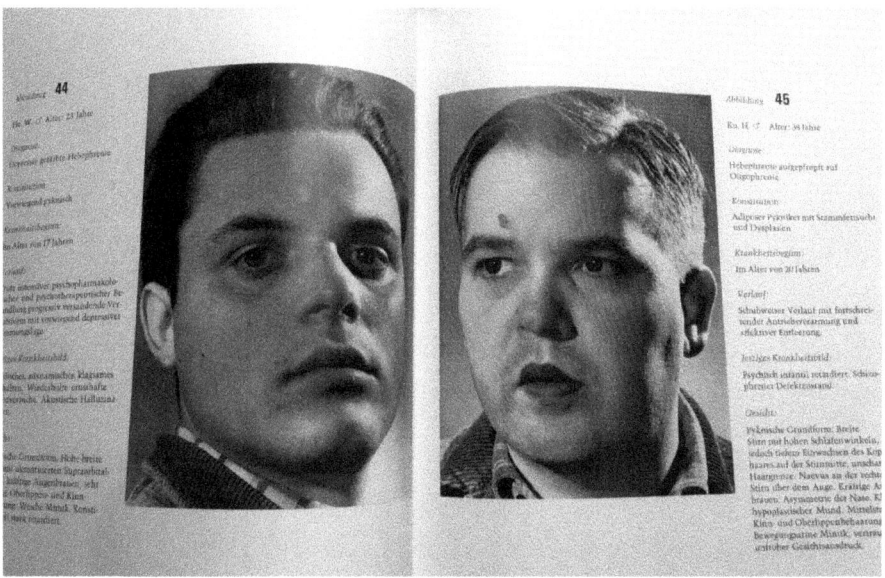

FIGURE 31 Pages from a post war text book produced by psychiatrist Gerhart Mall
The textbook depicts patients with 'unusual' syndromal clinical presentations.

Nitsche was arrested on 3 November 1945 and taken into custody in Dresden in the Soviet occupation zone. In what became known as the 'Dresden trial' (16 June–7 July 1947) Nitsche was one of 15 defendants charged with 'crimes against humanity'[354] perpetrated at Pirna-Sonnenstein and the hospitals at Großschweidnitz and Leipzig-Dösen. The charges against him included his leadership role in the T4 administration, his direction to nursing staff at Großschweidnitz and Leipzig-Dösen to administer dangerous doses of Luminal, and the actual killings using the *Luminalschema*[355]. Nitsche's first lawyer resigned as he found his client's actions 'against his own beliefs'.[356] His second lawyer sought to base Nitsche's defence on the findings of Ewald Meltzer's parent survey about the acceptability of euthanasia (as discussed in chapter 2) as well as the mortality rates of asylum patients historically.[357] While the trial was conducted under German law by German jurists, the occupying Soviets monitored proceedings closely. Recent research has investigated Soviet influence in the proceedings and found that the trial could not be considered a Soviet show trial.[358]

Even though there is no transcript of evidence for the Dresden trial, the records of interviews and interrogations of Nitsche provide an insight into what he professed were his motivations. Nitsche had argued that in the case of 'very damaged' patients, euthanasia was the only option. He claimed that when he reviewed files of patients killed in the various sites in the *Krankenmorde*, he was convinced by examining many of the photos of the victims that this was the correct course of action.[359] Nitsche sought to distance himself from the motivations of Brack and other functionaries in the KdF, stating that he rejected the arguments of killing patients on the basis of their minimal work capacity. Nitsche claimed that for him 'it was really the perspective of the patient that this was a *gnadentod* [mercy death]…ending a life that was a torture for the patient and the family'.[360] After he was found guilty of all charges,

Part 1 - The Crime

Nitsche's lawyer maintained that his client's motivations were entirely benevolent, specifically that *Gnadentod* was an act of mercy and that if Nitsche was guilty of crimes against humanity, then countless other philosophers, jurists, academics and physicians were also guilty. He averred that *Gnadentod* did not automatically become a 'crime' because it was advocated by the Nazi regime.[361]

FIGURE 32 The Dresden 'Doctors' trial', 1947—1948
Nitsche is far right in the dock.

Hermann Paul Nitsche was convicted of all charges and executed on 25 March 1948 by guillotine in the courtyard of The Royal Court of Justice at Münchner Platz, Dresden.[362] As with the records of victims of his *Aktion* T4 medical committee deliberations, the official notification of Nitsche being put to death includes a '+' marked in red pencil against his name.[363]

PART 2
LESSONS AND LEGACIES

'Whatever proportions these crimes finally assumed, it became evident to all who investigated them that they had started from small beginnings. The beginnings at first were merely a subtle shift in emphasis in the basic attitude of the physicians. It started with the acceptance of the attitude, basic in the euthanasia movement, that there is such a thing as life not worthy to be lived. … it is important to realize that the infinitely small wedged-in lever from which this entire trend of mind received its impetus was the attitude toward the nonrehabilitable sick'.

Leo Alexander (1949) *'Medical Science Under Dictatorship'*,
New England Journal of Medicine, Volume 241.

PART 2
INTRODUCTION

Many parts of the story of Elvira and Lisa Hempel are unique to their time, although the tale has a universal resonance. In reading this book one is confronted with the social exclusion, maltreatment, and persecution of a group of people deemed 'undesirable' or 'unworthy'. While enabled by a totalitarian regime possessed of a murderous agenda, these crimes against humanity were executed by trusted and powerful institutions and professions. One is left to ask why and how health and legal professionals, religious organisations and state institutions chose to abuse their power to first legitimate and then participate in the attempted annihilation of the most vulnerable people under their care. One also must ask why so many individuals either participated or were bystanders in these crimes, while others resisted, often at great personal risk?

The history of the *Krankenmorde* is littered with accounts of psychiatrists and other health professionals (mis)using positions of power to persecute violently and abuse people entrusted to their care. The Nazi rhetoric of the biomedical professions as benevolent protectors of a strong and healthy *Volk* belied involuntary sterilisation procedures that killed, injured and devastated the lives thousands of people. Many people compelled to sterilisation 'desperately resisted', but were 'forcibly taken to their procedure by the police.'³⁶⁴ Up to 5,000 people died from complications of enforced sterilisation procedures, while the hundreds of thousands who survived lived with the affront of this assault on their body and personhood. Far from the ideas of a 'mercy death' expounded in Nazi propaganda and in many of the perpetrators' legal defences in post-war prosecutions, so-called 'euthanasia' was a crime of intentional violence of which many perpetrators and their collaborators were aware and was soon well known to victims. 'A great number of them knew of their peril. They wept, they begged not to have to go, they resisted. One [...] asked everyone for forgiveness, told our head nurse we will meet again in Heaven, and said to a Housefather, 'Our blood cries out for vengeance'.'³⁶⁵ People with mental illnesses, disabilities, or otherwise deemed 'unworthy of life' were starved, gassed, electrocuted, lethally drugged or died in the course of tests of other killing methods, after which their bodies were exploited or defiled before being disposed of in crematoria or mass graves.

These crimes were grounded in and empowered by biomedical science, which through various theoretical rationalisations and social policies, justified the social exclusion and maltreatment of people living with mental illnesses and disabilities or who were otherwise deemed 'socially undesirable'. These events highlight the power given to specialised forms of knowledge and their institutions, such as psychiatry and other disciplines in medicine. The *Krankenmorde* reminds us how such groups shape public knowledge and debate, giving power to professional experts who collaborate with other authorities (political, economic, legal, etc.) to construct and enact social programs for either benevolent or malicious intentions. In this case; a system determining who decided a person was 'incurable' or 'useless', what that meant to society, and what was to be 'done' about it. The imposition of these arbitrary and discriminatory categories of person formed an act of epistemic violence against people[366], an abuse of biomedical knowledge that underpinned the physical, social and political violence that ensued. The focus of psychiatrists on the utility of each patient and the pursuit of a 'cure' more than a 'care' highlights, paradoxically, the futility of many professional practices and state policies during this period.

The entire project of the Nazi state of mass elimination of the 'others'—through genocidal wars of aggression and the persecution and murder of many of its own citizens—has cast a shadow over the nation and its people ever since. The capital of the reborn *Bundesrepublik* is dotted with many memorials that warn and chasten Germans about their nation's crimes against humanity that are still in the realm of living memory. Despite many of the medical and nursing perpetrators evading justice, the reputation of the German health professions at home and abroad was tarnished for the decades that followed.

Our current approach to mental health, disability, social policy, human rights and ethics is illuminated by the moral shortcomings of health professionals empowered by the Nazi regime, who sought to degrade and ultimately destroy those under their care who they had failed to 'cure' or return to an acceptable degree of social and economic participation. The power endowed by the Nazi state to health professionals to eliminate these obvious failures of their discipline, emerged from a form of 'structural violence'. The term 'structural violence' refers to the 'social structures – economic, political, legal, religious, and cultural – that stop individuals, groups, and societies from reaching their full potential…[and] that put individuals and populations in harm's way'.[367] The structural violence of the medical and health professions in the Nazi period fostered and compounded the direct violence inflicted on those patients, whose complex problems could not be solved. The shadow of the *Krankenmorde* should prompt us to re-examine continually, contemporary health care practices and policies for evidence of structural and epistemic violence, the professional rhetoric that often accompanies proclaimed advances in biomedical science (particularly in psychiatry), and the political, economic and social power of these activities.

The collusion of the biomedical and health professions in these crimes also urges one to consider the moral dimensions of science and the danger in assuming that biomedical research, technologies and practices evolve in a moral vacuum and are free of values or ethical constraint. The aftermath of the *Krankenmorde* further compels one to consider ideas about guilt, responsibility, and atonement—both the social and political contexts of these kinds of deliberations and the broader lessons for professions and other groups in 'working the past'.

The series of events that began with the murder of the infant Gerhard Kretschmer in 1939 and ended in the *Reinhard* extermination camps in 1942-43, prompt one to consider the nature and extent of the Holocaust; the relationship between exclusion, persecution and annihilation on grounds of both race and biology; and how societies progress toward such crimes. In the following chapters we will situate the *Krankenmorde* in the broader history of the Holocaust and in relation to the concept of genocide; unpack the significance of these events as they relate to contemporary issues in biomedical ethics; reflect upon questions of how as a culture we commemorate, remember and atone for such a collapse of professional morality and humanity; and consider the life of a survivor of the *Krankenmorde*.

CHAPTER 7
WAS ELVIRA HEMPEL A SURVIVOR OF THE HOLOCAUST?

Following the morally complex and emotionally disturbing work that culminated in the Nuremberg Doctors' Trial of 1947, Leo Alexander returned to the United States and took up a professorship in the Tufts Medical School in Massachusetts. Deeply affected by what he had encountered in Germany, Alexander reviewed those events and embarked on a series of considered responses. One publication, released in 1948, proposed the introduction to the medical lexicon of the term 'ktenology' to describe the study of mass killing. In coining the neologism, Alexander had appropriated the ancient Greek term '*apokteinō*' meaning 'to kill'.[368] In this he sought to define what he had seen of the crimes of the German and Austrian medical professions who had colluded with the Nazi regime to develop more efficient means of killing. Alexander's more famous publication on the Nazi era was a 32 page monograph titled 'Medical Science under Dictatorship'[369] which was published in 1949 in an abbreviated form in the *New England Journal of Medicine*.[370] In this 1949 paper, he also drew a connection between the *Krankenmorde* and the Holocaust, noting the 'euthanasia' program was 'merely the entering wedge for exterminations of far greater scope in the political program for genocide of conquered nations and the racially unwanted. The methods used and personnel trained in the killing centres for the chronically sick became the nucleus of the much larger centers on the East, where the plan was to kill all Jews and Poles and to cut down the Russian population by 30,000,000'.[371]

The idea that the *Krankenmorde* was both a precondition and preliminary step to the Nazi's 'Final Solution' became an integral part in the history of the Holocaust, particularly in the landmark works of Henry Friedlander and Robert J Lifton.[372] This connection between those two crimes begs the question, were the victims of the *Krankenmorde* also victims of the Holocaust? As discussed previously, Jewish patients in asylums and care institutions were placed in a different category from other patients under *Aktion* T4. Their murders within the T4 operation were primarily based on racial grounds and treated separately to the medical criteria applied by the T4 medical committee in Berlin. Arguably this makes these Jewish patients among the first victims of the Holocaust—their deaths were the initial step of what was later to become the 'Final Solution'. If this is to be accepted, in what other ways can we consider the *Krankenmorde* part of the Holocaust?

The 'Holocaust' is generally understood as the attempted destruction of Europe's Jewish population by the Nazi regime during the Second World War. The Holocaust has become totemic in Western history and culture, and Auschwitz—via the horrific crimes perpetrated in the concentration and 'extermination' camps located there—has become the primary cultural signifier of that historical moment.

The term 'Holocaust' dates from the thirteenth century, describing a 'burnt offering'.³⁷³ The custom of referring to the attempted genocide of European Jewry by the Nazi regime as 'the Holocaust' comes from the advocacy of Auschwitz survivor Elie Wiesel, invoking the near sacrifice of Isaac in the *Book of Genesis*.³⁷⁴ The term 'Shoah', meaning 'catastrophe' or 'destruction', applies specifically to the mass murder of Jewish people in Europe in ghettos, pogroms, concentration camps, *Einsatzgruppen* operations and extermination camps. However, the question of whether the term 'Holocaust' should apply to the persecution of all victim groups by the Nazi regime remains contested.³⁷⁵ According to Yad Vashem—the main Israeli institution for the commemoration of victims of the Nazi regime—the term 'Holocaust' refers to 'the sum total of all anti-Jewish actions carried out by the Nazi regime between 1933 and 1945: from stripping the German Jews of their legal and economic status in the 1930s; to the segregation and starvation of Jewish people in the various occupied countries; and the murder of close to six million Jews across Europe. The Holocaust is part of a broader aggregate of acts of oppression and murder of various ethnic and political groups in Europe by the Nazis'.³⁷⁶ Yad Vashem takes the position that 'Shoah' is the preferred term for the Nazi's crimes against the Jewish people as '(m)any understand Holocaust as a general term for the crimes and horrors perpetrated by the Nazis; others go even farther and use it to encompass other acts of mass murder as well'.³⁷⁷ In this formulation, the term 'Shoah' denotes the Nazis crimes against the Jewish people throughout this period of government.

The uniqueness and Judeo-centricity of the Holocaust is also subject to debate. Since the end of the Soviet Union (1992), the approach of historians to the Holocaust has moved between conceptualising it as a distinct form of genocide born of radical anti-Semitism, to it being categorised as one of many murderous episodes in the so-called 'Bloodlands' of Eastern Europe in the first half of the twentieth century[378]—and, often naively, equating Stalinist and Nazi crimes.[379, 380]

While controversial, the concept of the uniqueness of the *Shoah* is held resolutely, particularly among Jewish scholars, partly as a response to perceived attempts to diminish its status among other genocides.[381] The debate has become emotive and at times intensely *ad hominem*, creating what American Holocaust scholar Antony Polonsky calls 'the suffering Olympics'.[382] Any broadening of the category of Holocaust victims has been similarly controversial. As the Nazi persecution of Europe's Jewish population is the only clear attempt at extermination of an entire group during that time, referring to other victims of the Nazi regime as 'Holocaust victims' becomes an affront to many Jewish people.

Writing in anticipation of the annual 'Holocaust Remembrance Day' on 27 January 2017, the British writer Howard Jacobsen used the phrase 'Holocaust avidity' to refer to the 'species of competitiveness whose aim is to wrest the Holocaust from the Jews who, in this narrative, are presented as greedily insisting on an exclusivity of suffering'.[383] Jacobsen considered this a form of Holocaust denialism. In 1993 American historian Deborah Lipstadt published her influential work *Denying the Holocaust: The Growing Assault on Truth and Memory* in which she characterised overt Holocaust denial as a form of anti-Semitism.[384] Lipstadt's views of Holocaust denial re-emerged in a broadened form when she criticised the failure of then US President Donald Trump to specifically mention Jewish victims on the occasion of the 2017 Holocaust Remembrance Day. Writing in *The Atlantic*, Lipstadt stated that the 'de-Judaization of the Holocaust, as exemplified by the White House statement, is what I term softcore Holocaust denial'.[385]

Part 2 - Lessons and Legacies

The *Shoah* in Europe during the Second World War is, arguably, distinguished as the most 'modern'—as opposed to the most recent—of genocides. By the end of 1941, one third of Europe's Jewish population had been murdered, predominantly by gunshot, in a series of massacres perpetrated by SS and other paramilitary and police formations. From then, with the decision to implement the 'Final Solution', the process *in itself* of organised mass killing was to be the product of the industrialised ktenological progress of the Nazis and their medical collaborators. To effect this 'modern' and malignant process of mass killing, the apparatus of the modern industrialised state—its fake organisations and offices, its official committees and bureaucratic practices, its dubious research activities and biomedical assessments, its mass transportation of victims to an industrial scale killing solution, and the deception or forced collusion of its citizens—all needed to be aligned for the purpose.

While the term 'modernity' has multiple meanings in various settings, it is generally recognised as the period in Western history that followed the European Enlightenment and the French and American revolutions in the eighteenth century. The essence of 'modernity' is the triumph of human reason over superstition, the elaboration of understanding based on scientific methods of observation, and the application of these observations to questions about the universe. The ideal of the 'modern' state is a well-ordered society that functions within a balanced rule of law, possessed of stable and independent social institutions and bureaucracies capable of managing its large-scale operations, such as security, health care or education.

The Polish born philosopher Zygmunt Bauman conceptualised the Holocaust in terms of modernity.[386] Bauman argued that the Holocaust represented both a central moment in Jewish history and, more broadly, the 'hidden face of modernity'. In the late 1980s and early 1990s he published several books addressing the relationship between modernity, bureaucracy, rationality and social exclusion. He examined how modernity, in what he later called its 'solid' form, involved removing unknowns and uncertainties through the exertion of human control over nature. Such control was achievable through establishing hierarchical bureaucracies and enforcing rules and regulations as forms of control and categorisation to be applied to the population. These scientifically-based functions attempt to impose predictability, security and stability. Bauman observed that the modern state subjugated the individual for the benefit of society and enacted a process of reducing the community to a faceless population, becoming subject to bureaucratic control through powerful social institutions. A critical element in Bauman's formulation of the Holocaust was that it was in part enabled by the disconnect between different parts of the killing process—for example, transportation, health professions, infrastructure—working independently in silos. The objectification or dehumanisation of a specific target group within the community using medicalised metaphors, enabled the state to package (contain) the target (disease) in a particular way and move it through the silos, until the final killers could complete the task. Applying the lens of modernity to the *Krankenmorde* highlights how a eugenic-inspired and medically-justified bureaucratic operation led to a systematised process of identification of a group of victims, their exclusion from the rest of the community, their mass transportation to dedicated killing facilities, and a means of deception of those who might resist.

In this way neither the *Krankenmorde* nor the Shoah could have occurred unless state functions such as mass transport and infrastructure, state security services, administrative bureaucracies, the various professions, and a compliant press, were coordinated in the commission of the crimes.[387] The *Krankenmorde*, in particular its *Aktion* T4 program, was a specific instance of this form of 'modern' mass killing. As Leo Alexander and later writers would show, the categorisation of the *Krankenmorde* and the Shoah as modern mass killings has sustained the assumption that the *Krankenmorde* was a 'prologue' or 'trial' for the later attempted genocide of Europe's Jewish population. But there remains the question of whether this process of mass murder of the sick and disabled, as well as the elimination of Europe's Jewish population, was part of a deliberate and unified plan.

Distinct from the elimination of the sick and disabled from the German *Volk*, there has been considerable debate among historians over the origin of the Nazi regime's long-term intentions and plans for the complete annihilation of Europe's Jewish population. From this has arisen an 'intentionalist' and a 'functionalist' view of the Holocaust within German contemporary history, especially since the 1970s.[388] The intentionalist view, represented by Karl Dietrich Bracher and Eberhard Jäckel[389], has argued that the Holocaust was the result of a long-term master plan of Hitler, evident in his early writings and consistent in his actions over time, and thus sees a top down influence on the evolution of German genocidal anti-Semitism. In contrast, the functionalist view of Hans Mommsen and Martin Broszat considered the Holocaust as emerging from intense rivalry between factions within the Nazi regime. Broszat has argued that the Nazi state was not a unitary or coordinated phenomenon but rather a series of autonomous power structures struggling against each other for influence within the regime. These battles between rival organisations and agendas within

Hitler's immediate sphere were the main influence on the strategic and policy directions of the Nazi state.[390] Hitler, who according to Broszat was a 'weak dictator', was motivated primarily by the key political aims of both achieving a *Lebensraum* (living space) for the Germans in the east and the elimination of the Jews from German (and later European) society. Broszat takes the position that Hitler consistently privileged the views of the group who offered the most drastic or extreme proposals within his circle of influence. The most radical idea or plan always prevailed in influencing Hitler's decisions, a process Broszat called 'cumulative radicalization'.

Later historians have argued a more moderate functionalist view of the Holocaust, which sees that the failed attempt to expel Europe's Jewish population, to either Palestine or Madagascar for example, created an indirect pathway to their mass extermination.[391]

In his 1987 landmark study *Rassenhygiene, Nationalsozialismus, Euthanasie*, Hans-Walter Schmuhl outlined a 'cumulative radicalization' theory to explain the evolution of the *Krankenmorde* and its relationship with the Holocaust.[392] Schmuhl considered the Nazi 'euthanasia' program as the result of a process of cumulative radicalization caused by a charismatically legitimized, polycratic system of rule, which had chosen an inherently radical eugenic-racial theory as a basis for its policies. In Schmuhl's view 'euthanasia' was the preliminary stage of the Holocaust of the European Jews, connected by the extension of the *Krankenmorde* to the occupied territories in the east. This position has not gone unchallenged and a more contemporary view is that the *Krankenmorde* cannot credibly be conceptualised as pre-determined and methodical.[393] Rather, it reflects the arbitrary progression of a series of criminal acts perpetrated by ruthlessly ambitious health professionals and their accomplices, within the context of a chaotic totalitarian regime motivated by malignant applications of genetic and racial theories. The process utilised the apparatus of the modern state to perpetrate an industrialised killing process that would be later deployed on a much larger scale in Eastern Europe.

According to 'functionalist' views of the Nazi regime, the *Krankenmorde* was an aggregate of crimes perpetrated against people with disabilities in Germany and occupied territories, distinct in time and place. Among the first victims of the *Krankenmorde* were Jewish patients in asylums. As their murders were perpetrated by those who would go on to establish the *Aktion Reinhard* extermination camps, they are arguably the first victims of the final solution. From 1941 the *Aktion* T4 killers murdered concentration camp prisoners, many of them Jewish, in gas chambers in *Aktion* 14f13. In 1943 some of the *Aktion* T4 killers sought to establish *Polizeihaftlager* (a Police detention camp) in San Sabba for Jewish and other victims.

It is important to note that the application of processes that culminated in the *Krankenmorde* and the Shoah were identified and provided to the Nazi regime and then framed as biologically-based processes led by the health professions. As Australian genocide scholar Colin Tatz has observed, the biological theories that form the primary basis of genocide 'came from within the scientific, medical, and academic communities—not from without as a political imposition by totalitarian governments. In the twentieth century, the members of the 'doctorhood' that formulated, legitimised, and justified biological solutions to social and political problems not only thought, expounded, and wrote about their findings but also acted out their beliefs.'[394]

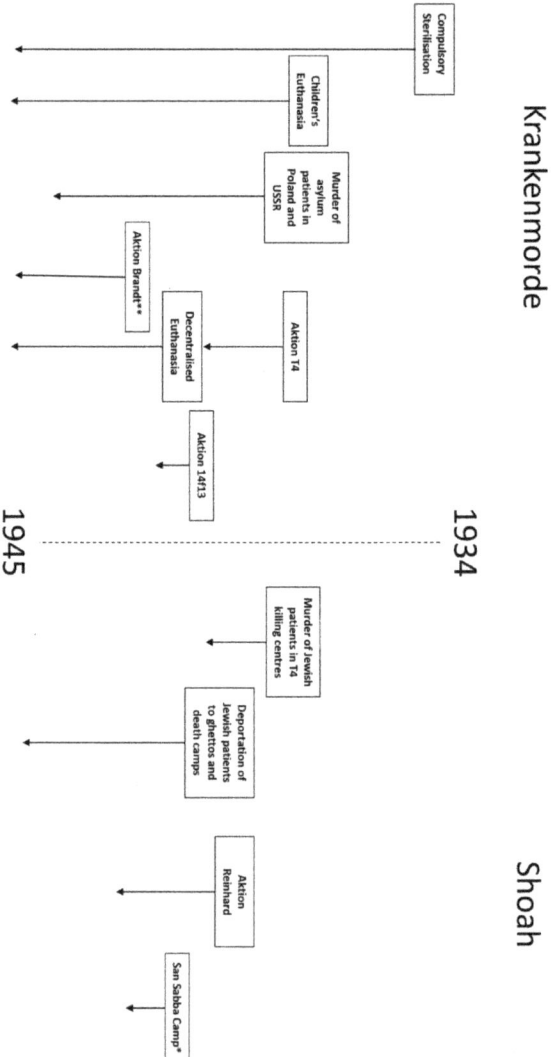

FIGURE 33 The historical relationship between the *Krankenmorde* and the Shoah

For comments on the San Sabba Camp see endnote 210, and for *Aktion Brandt* see endnote 332.

While many Jewish patients became victims of the *Krankenmorde*—and are, in Yad Vashem's terms, to be considered also as victims of the Holocaust and Shoah—the hundreds of thousands of non-Jewish victims of the *Krankenmorde* cannot be considered part of the Holocaust or the Shoah. What is clear is that all were trapped in an intentional process of industrialised mass murder, the operation of which presented a model that could be adapted and repeated in other settings by the Nazi regime to devastating effect.

Could the *Krankenmorde*, an act of premeditated-mass murder, then be seen as an attempt at genocide perpetrated against people with disabilities? If genocide was to be defined with reference to a perpetrator's intentional attempts to destroy an identifiable group of people— 'killing by category'[395]—then it would seem to fall into that broad meaning.

In 1944 Polish lawyer Raphael Lemkin coined the term 'genocide', arguing that 'The realities of European life in the years 1933-45 called for the creation of such a term and for the formulation of a legal concept of destruction of human groups'.[396] The word combines the antique terms '*genos*' (Greek: 'race' or 'tribe') and '*-cide*', (Latin *caedere*: 'extermination'). Lemkin defined genocide as 'a coordinated plan of different actions aiming at the destruction of essential foundations of the life of national groups, with the aim of annihilating the groups themselves'.[397] In 1948 the United Nations General Assembly adopted Lemkin's formulation in its Resolution 260A (III) Article 2 that defined 'genocide' as 'any of the following acts committed with intent to destroy, in whole or in part, a national, ethnical, racial or religious group, as such, killing members of the group; causing serious bodily or mental harm to members of the group; deliberately inflicting on the group conditions of life calculated to bring about its physical destruction in whole or in part; imposing measures intended to prevent births within the group; or forcibly transferring children of the group to another group'.[398] The crimes listed in the UN Convention all featured in the *Krankenmorde*. As the UN's construct of genocide does not acknowledge 'disability' or 'illness' as the defining characteristic of a targeted group, the *Krankenmorde* does not *prima facie* meet current international criteria for a genocide.[399]

Many in the field of disability studies have argued that 'disability' is as much a social category as religion, race or gender. In reviewing the history of the first part of the twentieth century in the United States or Europe, the socially distinct categories of 'disability' and 'race' can both be seen to have suffered from speculative and malign theories about biological inferiority and the perceived need for social control, exclusion and persecution. It is evident that the community living with disabilities under the Nazi regime represented a social category of biological 'inferior'; their identification and subsequent suffering and death conforms with the conventional definition of 'genocide' and their intentional destruction was facilitated by the apparatus of a modern state.[400] Whilst not as severe in prosecuting its controlling agenda (identify—isolate—eradicate), the United States and many other countries had embarked on a similarly biased and harmful trajectory.

However, since its legal and political inception, genocide has become a thorny and challenging issue in contemporary geopolitics and international relations. The refusal of the international community to assert the slaughter in Rwanda in the 1990s[401] and more recently in Syria and Myanmar as 'genocides'[402]—as against other forms of politicised mass killing that do not fit within the category—indicates that UN member states are seeking to avoid its application. Article 1 of the UN Convention on genocide compels signatories to 'prevent' and 'punish' genocide. Put into effect, this could involve military and economic interventions that may be contrary to national interests and impact on relations between signatory states. The use of the term is also an affront to the historic and nationalistic sensibility of some modern states. This is manifest in present-day Turkey's reaction to the accusation of 'genocide' in the deaths of more than 1.5 million Armenians, Greeks and Assyrians during the creation of the modern state in that country between 1915 and 1920.[403] There is also Australia's reluctance to acknowledge a colonial-era genocide of Aboriginal people, elements of which continued well into the twentieth century and, some argue, remain entrenched in the present.[404]

British scholar Martin Shaw has noted that in present-day geopolitics, the conceptualisation of genocide has become so narrowed that actual genocides are often deemed 'little more than mass murder, so that cases in which the majority of a population are not killed' are excluded from consideration. This leads to 'a proliferation of problematic new concepts like 'ethnic cleansing', 'politicide' and other '-cides' that describe aspects of what Lemkin, appropriately, saw as genocide'.[405]

Given the reluctance of the nation states' signatory to the UN genocide convention to actually countenance genocide in international affairs, and the propensity to reframe genocidal events as other categories of crimes against humanity, what is at stake by redefining the *Krankenmorde* as a genocidal act?

It is clear that the aim of the Nazi regime in the *Krankenmorde* was to eliminate those Germans and Austrians who, because of illness or disability, were unproductive and whose ongoing existence represented a drain on the population and whose reproduction was a threat to the health of the *Volk*. Unlike the Shoah, there was no decision taken to eliminate all people with mental illness or disability because they were mentally ill or disabled.

A critical issue in defining and prosecuting genocide is establishing the 'intention' of the perpetrator. Intentionality is evident in the persistence of a genocidal policy, regardless of any statement or documentation of the desire to eliminate the target group.[406] At the International Criminal Tribunal for the former Yugoslavia (ICTY), the prosecutors of former President of Yugoslavia Slobodan Milosevic struggled to demonstrate the intentionality of his alleged crime of genocide. Milosevic's death prior to the completion of the trial meant no actual verdict was reached on this issue.[407] Despite Hitler's written authorisation for provision of *Gnadentod* on specious medical grounds, there is no evidence of a stated intent by the Nazi state to eliminate all Germans and Austrians with disabilities. The sterilisation, persecution and murder of those patients who could not work or who had a poor clinical prognosis was relentless throughout the war, but this was not the attempted annihilation of all persons in the category of 'mentally ill' or 'disabled' as was the intention with Jewish victims. As such, the case for the *Krankenmorde* as a genocide is weak.

With the problematic legal application of the Genocide Convention since 1948, and extreme sensitivity among governments when confronted by it, there has been little fortitude within the international community to apply the Convention to contemporary acts that likely constitute genocide. There also appears to be little prospect of expanding the definition of genocide to include disability as a persecuted category. What then is to be gained by examining the *Krankenmorde* as an act of genocide?

The Shoah—the deliberate identification and murder of Jewish people in Europe—emerged from an extreme form of racism. The spectre of the Shoah functions as a motivator to present-day moral reflection on anything redolent of the racism that underpinned it. The *Krankenmorde* emerged from an extreme form of ableism. Ableism is 'a pervasive system of discrimination and exclusion that oppresses people who have mental, emotional, and physical disabilities'.[408] The ableist view of disability assumes that the totality of a person is their disability and that their defects or faults require mitigation to participate 'normally' in society. Institutionalised ableism is reflected in the failure to accommodate the needs of people living with disabilities; for example, the failure to ensure suitable access to buildings or the failure to apply effective strategies to accommodate learning difficulty or sensory impairment in education systems. While ableism is a contemporary concept, defining, diminishing and excluding people because of their physical, intellectual or psychosocial disabilities is not. Through eugenic discourse, ableism and racism were kindred phenomena during the middle of the twentieth century. The Nazi persecution of people on grounds of either race or disability emerged from the regime's creation of categories of biological 'Otherness', utilising scientific metaphors such as cancer, infection or genetic impurity to position such biological 'Others' as threats from within to the superiority and productivity of a 'master race'.[409] This enabled ableist and racist discrimination to become an acceptable narrative and driver for persecution throughout the state and community. In the present day they are analogous—as the Shoah compels us to reject racism, so should the *Krankenmorde* be seen for what it is and be used to reveal and revoke ableism.

If not a genocide, the *Krankenmorde* was an extreme act of ableism, an understanding of which has only recently entered historical consciousness. While the term and the full extension of its troubling social, physical and moral conception are not yet unpacked, there is much evidence to consider. The *Krankenmorde* is by no means the only instance of human rights violation to emerge from a society with an embedded history of ableism.

Just as the world was coming to terms with accounts of the horrors of the concentration camps and the graphic images of their liberated victims, the American public was confronted with their own group of emaciated and wretched victims of psychiatry. In May 1946, *Life Magazine* published a photo essay containing 36 disturbing images of patients in a psychiatric hospital under the title 'Bedlam 1946 - Most US Mental Hospitals are a Shame and a Disgrace'.[410] The story was based upon photographs taken secretly by a nurse, Charlie Lord, at the Byberry psychiatric hospital in Philadelphia. Lord was born in Iowa in 1920, raised a Quaker and, as a young adult, chose to become a missionary. Like many Quakers, he was a conscientious objector to the war and was drafted into the Civilian Public Service Unit. An alternative option was to volunteer as a research subject and be infected with Hepatitis A.[411] As part of his community service, Lord was assigned to work at the Byberry State Hospital in a neighbourhood of northeast Philadelphia. Like many others co-opted into working at Byberry, Lord was deeply troubled by a system that allowed emaciated and soiled patients to live in filthy conditions in the hospital's back-wards, and he sought to make their plight known.[412] Many Americans were outraged at the images published in the *Life Magazine* article, drawing comparisons with what had been revealed in newsreel footage of the liberation of concentration camp victims in Europe. Social commentator Albert Deutsch was damning in making a direct comparison between American psychiatric hospitals and *Aktion* T4 in his 1948 book *The Shame of the States*, claiming that 'we are not like the Nazis. We do not kill off "insane" people coldly as a matter of official state policy...We do it by neglect'.[413]

We have seen that such deeply troubling failures in psychiatry and mental health systems were not limited to any one nation. For many decades, numerous countries influenced each other as they followed a similar path. The progression of eugenics and euthanasia from an intellectual and ideological theory to a potential medical option to policies of sterilisation and then institutionalised murder was rapid and remains an enduring source of bewilderment and concern. Nor have they been they left behind as relics of the first half of the twenty-first century. In 2017, the UN Special Rapporteur on the right to health, Dainius Pūras, issued a scathing assessment of the state of global mental health care in the twenty-first century—including the dominance of the biomedical model of mental health and mental illness which, he argued, remains 'an untenable situation of unmet need and human rights violations'. Pūras, a child and adolescent psychiatrist from Lithuania, noted that with the UN adoption of various human rights instruments since the Second World War, there had been increasing attention paid to the question of human rights in mental health and psychiatry. Yet, with considerable concern he challenged this, posing 'whether the global community has actually learned from the painful past remains an open question'. [414]

In the same year, the UN Special Rapporteur on the rights of persons with disabilities highlighted the challenges for girls and young women with disabilities who are 'disproportionately subjected to forced and involuntary sterilization for different reasons, including eugenics, menstrual management and pregnancy prevention'.[415] The UN had recognised that the forced sterilisation of persons with disabilities 'constitutes discrimination, a form of violence, torture and other cruel, inhuman or degrading treatment, [however] the practice is still legal and applied in many countries'.[416] The report noted there are now more than one billion people in the world with psychiatric, physical and intellectual disabilities facing constant social disadvantage such as poverty and discriminatory laws and practices. Due to physical illness co-occurring with mental illness or intellectual disability, the life expectancy of this group of people is considerably shorter than the general population.[417] Nearly 80 per cent of men and women with serious mental illness who die prematurely do so as the result of poor physical health and a lack of access to healthcare[418], creating a situation described as a form of 'creeping euthanasia'[419], an echo of Deutsch's remarks some 70 years before. Within these revealing and disturbing observations come the questions of how we should value the lives of people living with chronic illness and disability, and how modern, decent governments and communities should provide equal access to appropriate care and services, and foster inclusive, non-discriminatory attitudes.

———

Whilst historians continue to research, assess and debate the complex relationship between the *Krankenmorde* and the Holocaust—the motivations for and enactment of deliberately harmful social policies and how these played out between supporter, contributor, resistance and victims; and ultimately what it reveals of society then and now—it is important to acknowledge that the Hempel sisters were victims of an extreme form of cultural oppression. This state sanctioned medical violence and murder, what we now see as an extreme form of ableism—intentional exclusion and persecution leading to murder—included people defined on the basis of disability. Accompanying the racism and anti-Semitism that served the malignant discourses of the Holocaust, the Shoah and the *Krankenmorde*, there was also ableism. Any clear reflection on the contemporary significance of the *Krankenmorde* will identify ableism as an underlying dilemma and motivator in the culture, public policy and political discourse of that society.

Elvira and Lisa Hempel were persecuted on the basis of a perceived disability; victimised by the medical, social, economic and educational forms of structural violence that had been enacted to fulfil the aims of the *Krankenmorde*. Here, the most extreme manifestation of biopower was at work. This form of violence, yet another expression of it under Adolf Hitler's Nazi Germany, underpinned the direct brutality they also endured.

We know little of Lisa Hempel's brief life other than she lived in circumstances of gross emotional abuse and neglect. It is probable that Lisa was affected by a global developmental delay arising from her deprived circumstances in a harsh and brutal institution. Her murder in Brandenburg in August 1940 was applied on the grounds of her presumed 'feeblemindedness'. We have no proof of the lies her family were told to cover up the crime, yet the weight of lies given to others provides a sense of that.

Elvira's incarceration and postponed death sentence were also based on her being diagnosed as 'feebleminded'—in retrospect on highly questionable grounds. This can now be seen as a medicalised justification for her persecution by the Nazi regime on the basis of her categorisation as a 'social undesirable'. While she avoided immediate death at the Brandenburg killing centre, her diagnosis would remain not just a medicalised and bureaucratic danger, but also a psychological burden—to be challenged continually by personal relationships, new settings and extreme social events. There was no 'surviving' such trauma until it had been understood. As was the case for so many survivors of the *Krankenmorde*, the Shoah and the Holocaust, survival meant remembering and questioning 'why was it so'?

CHAPTER 8
THE ETHICAL DIMENSION

On 11 January 1964, the US Surgeon General, Luther Terry, released a landmark document, *Smoking and Health: Report of the Advisory Committee to the Surgeon General of the United States.*[420] Terry and a panel of experts reported their findings that, among other things, cigarette smokers have a 70 per cent increase in age-corrected mortality rate; that there is a robust correlation between smoking, emphysema and heart disease; and there is a substantial causal link between cigarette smoking and lung cancer. Considered a watershed moment in public health, the Surgeon General's report was responsible for the subsequent strenuous efforts of all health authorities to reduce or ban cigarette smoking.[421] American historian Robert Proctor has since noted that tobacco is 'the deadliest artefact in the history of human civilization'. Smoking has caused the deaths of 100 million people in the twentieth century—more fatalities than was caused by both world wars combined.[422]

Luther Terry's revelations about the public health problems posed by tobacco were not new. Proctor had also observed that earlier in the twentieth century, German scientists had reviewed and concluded that cigarette smoking was a huge threat to both human health and the public purse. Much of that research, and the development of public health strategies to reduce smoking rates, occurred under the Nazi regime.[423] On the challenges posed by smoking and other public health hazards such as alcohol, low fibre bread and workplace safety, the Nazis were ahead of their time. Nazi public health policy saw interventions such as bans on smoking in workplaces and public venues, restrictions on tobacco advertising, and strict requirements for workplace safety that are not out of place in the present day.[424]

How does this seemingly progressive public health agenda of the Nazi state reconcile with the murder of Lisa Hempel and hundreds of thousands of others in the *Krankenmorde*?

———

Contemporary health professionals and scholars are wary about present-day comparisons with the Nazi period. The different positions taken in response to that era often reflect a disgust or revulsion—a violent accusation or emotional withdrawal—that denies a careful examination of social and medical policies and practices of that time. Around this lacuna lie complex, contested and varied beliefs about the uniqueness of the Holocaust and other Nazi crimes, and to what extent claims about current lessons from, or analogies to, the Nazi period are either accurate or appropriate. In making a case for the inclusion of Holocaust analysis in contemporary bioethics and medical education, some scholars argue that while the Holocaust was unique and not a balanced precedent for further bioethics deliberation, the motivations and value judgements of the perpetrators of medical crimes under National Socialism provide a critical perspective for contemporary bioethics.[425]

Simplistic comparisons to Nazis and the Holocaust in public discussions have become known as 'Godwin's Law'. Formulated by US lawyer Mike Godwin in the early 1990s to express his distaste for such glib comparisons, Godwin's Law maintains that 'as an online debate increases in length, it becomes inevitable that someone will eventually compare someone or something to Adolf Hitler or the Nazis'.[426] However, rather than considering such debate unworthy, Godwin adds that not all comparisons are inappropriate and the 'best way to prevent future holocausts, I believe, is not to forbear from Holocaust comparisons; instead, it's to make sure that those comparisons are meaningful and substantive.'[427]

Understanding the contemporary meaning and influence of the *Krankenmorde* has proven more challenging than many other instances of Nazi medical criminality, such as the medical experimentation crimes which have been central to the development of modern codes of research ethics. While the *Krankenmorde* has been less examined and understood than the significant other crimes of the Nazi regime, evolving knowledge about its events, its victims, its bystanders and its perpetrators has presented new opportunities to grapple with the complexity of its significance. Historian Michael Burleigh, for example, highlights the limitations of attaching to current arguments about euthanasia the immoral weight of the Nazi 'euthanasia program', while also acknowledging the usefulness of the historical analogy to some modern eugenic policies, particularly in reference to public policies in China and Singapore (discussed further below).[428]

In focusing on questions about medicine and healthcare, two overarching themes emerge from the *Krankenmorde* which may provide a useful focus for reflection, the first being the relationship between medical power and violence; the second is the tension around concepts of a 'better' or best life and a 'good' death. In recent years, the problematic notion of a 'good' life and death has re-emerged in public debate about euthanasia and here there are lessons from the *Krankenmorde* that could help guide moral and policy positions at the end of life. At the same time, directions in public policy, and commercial strategies behind assisted fertility and pre-natal genetic screening and manipulation, give rise to another emotive and perplexing arena for discussion. In the convergence of these discourses we see contemporary challenges that will benefit from a reappraisal of eugenics and its manifestations in the *Krankenmorde*. As people living with disabilities remain one of the most disadvantaged groups across all societies[429] this leads us to examine how we value and treat people living with mental illnesses and various forms of disability, while also reflecting seriously on how we have progressed since the *Krankenmorde*.

Here we will focus on the *Krankenmorde* via the lens of bioethics. Bioethics is a type of practical ethics concerned with such fundamental questions as 'How should I live?' and 'What ought we do?' applied to quandaries that emerge from the biological sciences, health care practice, and their related scientific, political, social and economic activities.[430]

From his research into the Nazi regime's public health policy, historian Robert Proctor has uncovered many elements that highlight the complexity of its apparently benevolent aims and challenge assumptions about its fundamentally malignant aims.[431] It is well known that the Nazi regime sought to create a racially homogenous 'Aryan' population –'*Volk*'– by removing all designated pollutants and impurities, whether heritable, racial, medical or environmental. The genetic and racial dimensions of this undertaking enabled the Nazi regime to use biomedical metaphors to justify a program of mass sterilisation and later extermination of these genetic and racial 'undesirables'. The Nazi regime's metaphor of Jewish people as either 'a cancer' or 'a bacillus' to be eliminated through extermination served its ideological and propaganda purpose, yet its concern about environmental and disease threats to the population was an equally significant driver of its public policies.

Proctor notes that before 1933 Germany had one of the world's highest rates of cancer. In the 1920s, German public health physicians had identified the likely causes of higher rates of cancer as being causally related to cigarette smoking, alcohol consumption, poor diet and environmental hazards. The Nazi regime had promulgated the metaphor of *Volkskörper* (people's body), conceptualising society as a healthy individual, a racial and genetically pure *Volk* comprising a 'racial state'. Public health initiatives to improve the overall health of the *Volkskörper*—and thus its productivity (and, ultimately, war-making capacity)— were promoted as individual health and lifestyle initiatives, a commitment connecting the individual to the state. This metaphor enabled direct intrusion into the lives of the citizens in the Nazi state in both the workplace and the home. By 1936 the Nazi regime had introduced stringent occupational safety regulations to reduce worker's exposure to radiation, asbestos and other workplace pollutants. This progressed to initiatives to encourage healthy lifestyles involving smoking cessation, improved diet, adequate physical activity and reduced alcohol consumption. Until the unfavourable turn of the war in 1942, the Nazi regime was selectively vigorous in its anti-tobacco initiatives, including implementing bans on cigarette advertising and smoking prohibition in work places or public spaces[432]—initiatives not out of place in present-day public health policy.

FIGURE 34 Nazi anti-tobacco propaganda from 1941
The caption reads 'He does not devour it [the cigarette], it devours him'.

How does one reconcile seemingly enlightened public health policy within a state apparatus that employed the gas chambers of Brandenburg killing centre and the starvation houses at Eglfing-Haar? Proctor concluded that this apparently benevolent approach of the Nazi regime in improving the health of the population was more an act of coercion and micromanagement of the lives of its citizens, much as the 1935 Nuremberg race laws dictated who a person could or could not marry. This was the prototypic 'nanny state' and the Nazi regime's appropriation of science as a means of exercising coercive power was as much an act of totalitarianism as the imprisonment of 'social undesirables' and 'enemies of the state' in concentration camps.

Recalling the concept of 'modernity' that was central to Zygmunt Bauman's framing of the Holocaust: a major function of modernity was human control over the natural world, including individual biology and public health. To Bauman, one of the distinguishing features of the Holocaust was that it was the culmination of the many activities that characterised modernity.[433] At the core of this exercise of power over biology by the Nazi regime was a range of activities from banning smoking or encouraging exercise and eating wholemeal bread, through to sterilisation and 'euthanasia' of biological undesirables.

The French philosopher Michel Foucault observed that the evolution of the modern nation state in the eighteenth and nineteenth centuries resulted in populations ruled by governments rather than by sovereign or ruling classes.[434] The modern 'state' was a codified set of power relations at all levels, and Foucault coined the term 'governmentality' to describe a way of administering populations in this new modern European nation state. In this newly formed social aggregate, methods of government bureaucracy and executive power evolved with control over the population expressed through government sanctioned institutions. In addition to raising taxes, administration of the rule of law and other functions of the state, the power exercised by governments expanded to managing the births, deaths, reproduction, health and illnesses of a population. In this way—with the recognised importance of the medical and biological sciences, and their control over the health of the individual and the nation—the executive power of the state was able to take over and directly influence funding and policies in that area. Foucault named this practice 'biopower'.

Foucault posited another manifestation of political and institutional power as control over knowledge existing in the form of a 'discourse': a specific way of speaking or writing about reality within a group (a 'discursive formation'). By defining what can be thought and said about the world and what cannot, discourse creates rather than discovers a form of truth. Foucault's discursive groups were often professional disciplines (such as psychiatry or public health) speaking authoritatively through a particular institution of state. Through this dominating source of authority, Foucault could show that control over knowledge—its construction, development, application and distribution—was indistinguishable from political power in the modern state.[435]

Discourses over health and illness—genetic or racial hygiene, public health policy or other means of controlling the biology of the population—were the essence of the Nazi form of biopower.[436] The state policies of sterilisation, exclusion, starvation, deportation, ghettoisation and ultimately genocidal elimination, were all forms of biopower in Nazi Germany. Genocide—the deliberate destruction of a group defined in terms of social or biological categories such as race (which could include, as we show here, disability)—is an exercise of biopower. In this instance, the apparatus of the Nazi state organised itself around advancing the life and well-being of some, while arguing for the exclusion and destruction of others, based on claims they compromised the life of the protected group. In this way the preferred metaphors of the Nazi regime for depicting Jewish people was to categorise them as cancer cells or bacilli infecting the *Volkskörper*. As we have seen, this same discourse was applied to those considered genetically inferior and mentally or physically deficient—people with disabilities. Exterminating them was tantamount to genocide, and the idea of genocide as a manifestation of 'biopower' is what Foucault and later writers such as Giorgio Agamben termed 'thanatopolitics'—the politics of death.[437]

In this view, the diagnosis of 'feeblemindedness' by physicians was ultimately a malevolent act as the label often had devastating consequences for the subject. In this context, the act of diagnosis was an example of what has been termed 'epistemic violence'—a form of violence perpetrated through the construction of a form of knowledge. The French sociologist Pierre Bourdieu had, in the late twentieth century, defined the phenomenon of 'symbolic violence' as the imposition of ways of perceiving and thinking about the social world by an empowered group in society. This imposed socially constructed reality then dominates the society and serves to maintain the power relationships between groups.[438] Symbolic violence often functions to create and maintain an 'Otherness' to those subject to the power imbalances created. Epistemic violence extends this concept closer to interpersonal violence, in that it has a subject, an object and an action. The violent manifestations of knowledge are frequently transacted by an empowered person upon another, creating an 'Otherness' in the person or group that has detrimental impact.[439] In his *History of Madness*, Foucault noted a profoundly harmful instance of epistemic violence in the redefinition of 'madness' as irrationality, which enabled the mass incarceration of people with mental illness in what he described as 'the Great Confinement'.[440]

The diagnosis of 'feeblemindedness' or other forms of disorder and disability enabled the most malignant of consequences for those it cast into a category of undesirable otherness: exclusion, sterilisation or death. Epistemic violence remains a form of little acknowledged biomedical harm, both through the profound influence of psychiatric diagnosis or categorisation of individuals through IQ testing. Psychiatric labels continue to potentially exclude, marginalise, stigmatise and discredit those subjected to their application.[441]

Present day expressions of biopower and epistemic violence as a means of state control include policies and strategies of health promotion focused on both individual and population levels. Programs screening for disease at all points of the lifespan, public policy initiatives to encourage healthy lifestyle and dietary habits, vaccination to promote 'herd immunity' against infectious diseases, and the funding and regulation of reproductive technologies are all manifestations of contemporary biopower. In a well-ordered liberal society these are usually effected by incentive and consensus rather than coercion. However, as market forces have come to influence health care, biopower has also become a tool of the corporate sector. Pharmaceuticals, medical devices, nutritional supplements and a burgeoning fitness and wellness industry all rely upon, and seek to influence, the exercise of biopower. In addition to powerful direct political and financial influence on public policy, corporations also assert their biopower through soft measures—from advertising and product placement in television, film and social media, through to more overt attempts to define ideal life and social acceptance via manipulation of the physical embodiment of celebrity. Popular television programs that feature a range of people who are obese and 'fat shamed' into attempting to lose weight, 'celebrity' marketed dietary consumer products or commodified fitness programs, are apposite examples of market-based biopower.[442]

The importance of understanding the *Krankenmorde* in the context of biopower, and vice versa, is not 'Godwining' in seeking to equate contemporary challenges with Nazi eugenic policies. Nor does biopower provide the only credible way of framing our understanding of the *Krankenmorde*. Rather, it helps conceptualise how a past (and any future) state can harness biopower and to what ends. We can ask questions such as 'is public vaccination of young girls against Human Papilloma Virus a biopolitical step, and to what end?' or 'is a tax on sugar a form of biopower, and who is advantaged or disadvantaged by this?' or 'to what extent is biopower being used to privilege certain forms of population selection and what can be learned from that?'. In this sense, biopower is a useful analytical framework for scrutinising developments concerning health, life and death and provides another way to think critically in bioethics. These forms of analysis help us to be mindful about how knowledge is shaped by power, who is invested in and applying that power, and how this can determine who or what is included or excluded, normal or abnormal, right or wrong, and in certain cases who lives or dies.

Part 2 - Lessons and Legacies

Biopolitics figures significantly in the complex debates around end-of-life decisions and assisted dying. The perpetrators of the *Krankenmorde* sought to legitimate and later defend their crimes by defining their actions in terms of 'mercy death', 'medical futility' and 'relief of suffering'. While these arguments remain the basis of the present-day case for legislated assisted dying, in the setting of the *Krankenmorde* they were a significant component of the deceit utilised in a state-controlled program of mass murder of parts of its own population. How then should the historical example of the *Krankenmorde* figure in our current deliberations on assisted dying?

In 2009, the term 'death panel' appeared in American political discourse as a calculated response to proposed health care reform legislation. The Obama administration's health care reform bill (*The Patient Protection and Affordable Care Act*) included a provision to authorise Medicare to reimburse doctors for appointments with patients for the specific purpose of discussing their wishes surrounding end-of-life care. The provision for advanced care planning in the legislation had 'widespread bipartisan support until the summer of 2009 when (former Governor of Alaska) Sarah Palin morphed talk of providing support for patients who wanted to have a discussion with their physicians concerning their priorities into rhetoric alluding to 'death panels".[443] In a 'Statement on the Current Health Care Debate', Palin wrote: 'The America I know and love is not one in which my parents or my baby with Down Syndrome will have to stand in front of Obama's 'death panel' so his bureaucrats can decide, based on a subjective judgment of their 'level of productivity in society,' whether they are worthy of health care. Such a system is downright evil.'[444] Political activist Lyndon LaRouche wrote of 'Obama's bill as a revival of the Nazi

T-4 euthanasia and genocide policy'.[445] Notwithstanding the gross misrepresentation of policies concerning end-of-life care planning[446], such canting highlights the intensity and complexity of contemporary debates about 'euthanasia' and the perils of drawing emotive analogies with the *Krankenmorde*.

Physician and academic Mary Tinetti argued that the potency of the 'death panel' canard had immediate and longer-term consequences. Political pragmatism from the Obama administration saw the advance care planning provision removed from the *Affordable Care Act*, compromising both the integrity of doctor-patient relationships and the legitimacy of end-of-life care planning.[447] That this misleading 'death panel' narrative could take such powerful hold in the vexed political debate over health care in the United States highlights the contentiousness of ongoing community deliberation about euthanasia—therein lie the perils of engagement with the *Krankenmorde* precedent. Yet, to simply accept that the murder of patients in the gas chambers and starvation houses of the *Krankenmorde* is in some way morally equivalent to decisions about end of life treatment by a patient suffering intractable cancer pain, is as ignorant and culpable as accepting the 'mercy death' euphemism proffered by the Nazi regime.

A deeper analysis of both the apparent motivations of the medical perpetrators of the *Krankenmorde* and end-of-life ethical dilemmas raises fundamental questions about the role of the medical profession and its perceived duties to value and protect life, and its obligations in the relief of suffering.[448] These fundamental concepts in medical ethics can and do come into conflict in the setting of end-of-life decision making. Within Sarah Palin's political critique of *The Affordable Care Act* was the deeper concern that faceless Government officials would determine the conditions for end-of-life. While a thorough analysis of the legislation and useful concepts such as biopower could have alleviated that dilemma, end-of-life choices—as we shall see—retain a range of emotive and challenging interpretations.

It is worth reflecting on some of the language and justification used by the proponents of the *Krankenmorde* as a means of understanding the medical value system from which it emerged. In 1947 Karl Brandt stated in evidence in the Nuremberg Medical Trial (NMT): 'Would you believe that it was a pleasure to me to receive the order to start euthanasia? For fifteen years I had laboured at the sickbed and every patient was to me like a brother, every sick child I worried about as if it had been my own. And then that hard fate hit me. Is that guilt? Was it not my first thought to limit the scope of euthanasia? Did I not, the moment I was included, try to find a limit as well as finding a cure for the incurable? Were not the professors of the Universities there? Who could there be who was more qualified?'.[449] Here Brandt seems to argue that the provision of a 'mercy death' was a benevolent act in the face of 'incurable' conditions.

In an interview conducted prior to his trial, Brandt had informed Leo Alexander that the 'euthanasia' program he coordinated applied to: 'terminal cases. The things which I am [charged with], the documents which are with the prosecution somewhere, all this does not correspond to those things [that] we were interested in. It was the pure idea of euthanasia, and in fact seen from the medical perspective. Therapeutic measures which are known and achievable could not help these patients any more. It was therefore also not possible to achieve a standstill in the development of the disease. The condition itself was agonising. That is a term "euthanasia" to deliver these humans from this condition'.[450]

In some ways Brandt's argument connects to the primary consideration underlying the permission to end life: to end suffering or to truncate a futile and painful situation in which no improvement is realistically to be expected. On this point Brandt and others succeeded in persuading the NMT judges. However, as the British academic Emmeline Burdett has argued, victims of the *Krankenmorde* were also victims of the NMT judges. In citing transcripts of the final judgement in the 'Doctors' trial', Burdett concludes that the NMT judges took the view that the state had the right to end the lives of those deemed appropriate for 'euthanasia' and that it did not constitute a crime unless the killings were based on racial grounds or perpetrated against citizens of other nations.[451]

In both the arguments of Brandt and present-day euthanasia advocates, considerations of futility and intractable suffering are described as the primary motivation towards 'euthanasia'. The critical difference between Brandt's argument and those offered in many contemporary debates in support of euthanasia is that the latter prescribe a situation of informed consent by a competent adult in a circumstance of, presumably, legally regulated, physician assisted, suicide. Brandt sought to defend the termination of life of diminished value without the person's direct or proxy consent. It is doubtful and unlikely that any of the victims of the *Krankenmorde* chose to end their lives in the state-administered killing program, any more than anyone chose to be shot dead in a pit in Ukraine during an *Einsatzgruppen Aktion*.

Brandt's view was not uniformly held among German physicians of that period. During the trial of Paul Nitsche in Dresden in 1947 (see Chapter 6), Richard Pfeifer, a neurologist from the University in Leipzig, provided an ethical commentary for the prosecution on the actions of Nitsche and his co-accused. Pfeifer stated unequivocally the opinion that physicians must only preserve life and not participate in executions or euthanasia. He argued that the limits of medical knowledge often made a lie of diagnostic or prognostic statements and that even the most impaired patient must be cared for—at this point Pfeifer used the analogy of feeding a useless old horse or dog with '*Gnadenbrot*' (mercy feeding).[452]

FIGURE 35   Karl Brandt at the Nuremburg Doctors' trial, 1947—1948
As the image was taken, Brandt was being sentenced to death by hanging by the US War Crimes Tribunal.

This may be jarring to contemporary sensibilities, yet at the time sought to communicate a notion of beneficence. Principles of beneficence underlie current formulations of medical ethics which emphasise the prohibition of the destruction of life, whatever value is placed upon it.

The contemporary conception of 'euthanasia' is generally described as the intentional ending of another person's life by direct intervention of a physician through the lethal administration of drugs at that person's competent and voluntary request.[453] There is a range of terminology describing different circumstances where life is ended, usually by administration of a lethal combination of medications, or where life-prolonging treatment is withheld. In his landmark work *Practical Ethics*, the philosopher Peter Singer argues, 'Voluntary euthanasia occurs only when, to the best of medical knowledge, a person is suffering from an incurable and painful or extremely distressing condition. In these circumstances one cannot say that to choose to die quickly is obviously irrational'.[454] There are also passive forms of euthanasia which involve the withholding of life-saving or life-prolonging interventions. This is either voluntary passive euthanasia, based on the consent of a competent patient or through an advance directive, or non-voluntary passive euthanasia, where a decision is made without the patient's consent to withhold such treatment (for example, in circumstances of severe brain damage). Involuntary euthanasia refers to circumstances where a patient can refuse euthanasia but is neither asked nor their refusal accepted. Physician-assisted suicide, in contrast, is defined as a physician intentionally assisting a person to end their life by the provision of drugs for self-administration at that person's competent and voluntary request. These approaches are considerably different to the biased processes and ideological system that delivered the *Krankenmorde*, where there was little to separate involuntary euthanasia from mass murder.

Aside from Godwin's Law, citing the *Krankenmorde* as an historical precedent in order to oppose legalised euthanasia can lead to what ethicists refer to as the 'slippery slope' argument: that if we allow position 'A' to come about, then it is inevitable that through some direct or indirect connection, position 'Z' will happen. The 'slippery slope' argument is a controversial debate.[455] In the minds of some, the 'slippery slope' argument is a fallacy in that it assumes, not always justifiably, that there is an inexorable progression to a morally reprehensible outcome through the gradual progression of seemingly innocuous steps.[456]

Arguments against any form of active euthanasia, such as those championed by American oncologist and bioethicist Ezekiel Emanuel[457] are based on the notion that better palliative care can obviate the need for assisted dying, and that any active intervention by a physician to end life will embark on 'the slippery slope', leading to a broadening of the grounds of such decisions and potentially including equivocal situations such as chronic pain or psychological distress.[458] The risks from falsely assuming long-term political stability and benevolence in public policy are also cited in arguments against euthanasia—where a change in government may lead to unintended and draconian applications of the euthanasia law.[459] While Sarah Palin's arguments in this setting were implausible due to their extremity, on the issue of the future abuses of what were originally well-intentioned euthanasia practices, such an argument is more defensible.

In support of qualified application of euthanasia, the views of philosopher and pro-euthanasia advocate Peter Singer have been at times compared unfavourably with the justifications used by the Nazi regime in its 'euthanasia' program. This has made Singer the subject of full-throated protests in Germany and extensive criticisms in some parts of the media and among his colleagues.[460] Singer argues that some forms of human life (for example, extremely premature babies, critically ill adults, or people living with profound disabilities) are less able to benefit from, and therefore less entitled to, limited social resources. These claims are strongly utilitarian and seek to justify euthanasia of adults and children with severe disability. In the case of the infanticide of a disabled baby, Singer's justification is that such an act enables the family to have a healthy child and increase the total happiness or preference gratification, the ultimate arbiter of utilitarian arguments.[461] Such an argument echoes the decision of Richard Kretschmer to request the killing of his son in 1939. In 2017, as some Australian jurisdictions were debating euthanasia legislation, Singer returned to the argument, drawing on the lived experience of legalised euthanasia that had followed his controversial writings of the 1980s. Citing the Godwin-esque slippery slope arguments against his position from decades before, Singer wrote 'this scenario never seemed plausible to me, but in the 1970s it was not easy to definitively refute it…Forty years on we have a much better basis for saying that allowing doctors to act on considered requests from their terminally or incurably ill patients will not take us down to places we do not choose to go'.[462]

Since 2002 a growing number of jurisdictions have legalised euthanasia and assisted dying, providing another means for examining historical experience. Laws with provisions for voluntary assisted euthanasia/suicide are now in place in the Netherlands (since 2002), Belgium (2002), Luxembourg (2009), and Canada (2016). Colombian law permits euthanasia (2015); while Switzerland and some states in the United States allow PAS (Oregon 1997; Washington 2009; Montana (via case law) 2009; Vermont 2013; California 2015; District of Colombia 2016; Colorado 2016)[463]. The Victorian parliament in Australia passed a voluntary assisted dying bill in late 2017.[464]

As Singer had inferred, recent official reporting and research into both the lived experience of euthanasia and the assisted dying laws in several jurisdictions have provided useful data for consideration. Some analyses suggest that euthanasia and physician-assisted suicide account for between 0.3 and 4.6 per cent of all reported deaths; that the frequency of these deaths slowly increased each year after the introduction of assisted dying legalisation; that most cases of assisted dying involve a patient with terminal cancer; that intractable pain is seldom the primary motivation for seeking assisted dying; and that the most common motivations for seeking assisted dying are the person's sense of loss of autonomy, dignity and other forms of psychological distress. Most significantly, there is no evidence suggesting abuse of assisted dying laws.[465] There is also little to indicate divestment in palliative care in jurisdictions in which euthanasia laws are implemented.[466] In 2004 the Dutch parliament introduced the 'Groningen Protocol' that permits the euthanasia of children with severe disabilities, although the rates of late-term feticide (abortion) and legalised infanticide in the Netherlands dropped following the introduction of the law.[467] Following amendments to the Belgium law in 2014, the first reported case of paediatric euthanasia in that country occurred in September 2016.

While the findings are generally cautious—and in some degree reassuring to those anxious about the 'slippery slope' where assisted deaths of patients with terminal cancer is concerned—the situation in Belgium, Netherlands and Luxembourg (the Benelux countries) has evolved in ways that have caused unrest. Assisted dying for people who are not terminally ill, such as those suffering from psychiatric illness or early stage dementia is legal in the Benelux countries.[468] Approximately 3 per cent of Dutch and 1 per cent of Belgian assisted deaths since 2002 were undertaken due to 'intractable psychiatric disorders.'[469] The laws in Benelux countries consider 'intractable suffering'—presumed to be the result of severe treatment refractory depression—as being reasonable grounds for assisted dying. Yet the available data indicates that depression was the main clinical problem in just over half of those patients who proceeded to euthanasia on psychiatric grounds. A significant number of these mental illness-justified euthanasia deaths involved patients diagnosed with personality disorders or autism spectrum disorders. Of particular concern is that one in five of these psychiatric euthanasia deaths were patients who had never been hospitalised. In many of these cases of euthanasia on psychiatric grounds, the clinical justification for euthanasia cited the person's 'social isolation' or 'loneliness'.[470] At the time of writing, Dutch law makers were considering modifying the euthanasia law to allow older people to seek assisted dying on the grounds of 'being tired of life'.[471] These developments mark a significant difference in approach between European countries with legislated euthanasia to similar laws in the United States.

In contrast to the growing enthusiasm in the Benelux countries for expanding the availability of euthanasia to non-terminally ill people with complex psychosocial problems, the prospect is anathema to nearly every national and international professional psychiatric organisation and has become a cause celebre within the World Psychiatric Association.[472] The World Medical Association (WMA) has long held that deliberately contributing to the end of a patient's life, even with consent, is unethical.[473] In October 2017, following the legalisation of physician assisted suicide in the Australian state of Victoria, the WMA stated that such a law created 'a situation of direct conflict with physicians' ethical obligations to patients and will harm the 'ethical tone' of the profession. It also warns that vulnerable people will be placed at risk of abuse and that a 'precedent will be set that physician assisted suicide and euthanasia are ethically acceptable'.[474] Assisted deaths determined on grounds of relative value judgements about the kind of social existence the patient has, and the potential normalisation of physician-facilitated death of people with mental illness on grounds of treatment refractoriness or medical futility, are a significant shift in the implementation of assisted dying legislation.[475] Setting aside the 'slippery slope' argument, there are significant ethical concerns at the death by euthanasia of a socially isolated and chronically depressed person with an ostensibly non-fatal illness in Amsterdam or Brussels.

Legalised euthanasia could be argued to legitimate the view that the lives of people who are elderly or living with chronic illness or disability are of diminished value.[476] This point was made eloquently by British actress and disability rights advocate Liz Carr[477]: 'If I said I wanted to die, the press, celebrities and the public would support my choice, seeing it as rational and understandable. Hell, they would probably set up a 'Go-Fund-Me' campaign to help me make it happen…Yet when a healthy, non-disabled person wants to kill themself [sic] it's seen as a tragedy, and support and prevention tools are provided. If nothing else convinces me that to legalise assisted suicide is not a safe option for many of us then this does. Suicide is not seen as socially desirable – so why is assisted suicide seen as compassionate when it's for ill or disabled people?'.[478]

The prospect of disability as a legitimate pretext to euthanasia is a major theme in disability rights discourses. A current dominant theme seems to frame the core existential dilemma facing a person living with a disability as a Hamlet-like deliberation between a 'bad life' or 'good death'. The 2016 UK film 'Me Before You' depicts a man living with a high level spinal cord injury deciding to seek assisted dying to spare his caregiver (and love interest) the burden of caring for him. He reframes his dilemma as a moral choice in allowing her to live a 'full life' without him rather than a 'half-life' caring for him. Films depicting people with disabilities often default to the question of euthanasia, a seeming homage to the central premise of *Ich klage an* (see Chapter 2). Director Clint

Eastwood's 2004 film *Million Dollar Baby* advocates the validity of euthanasia in the circumstance of a high level spinal injury and was extraordinarily successful in that year's Academy Awards. Both *Ich klage an* and *Million Dollar Baby* depict the moral arc travelled by a paternalistic male protagonist, from his initial horrified rejection of the notion of facilitating the mercy death of a previously talented and beautiful woman with a horrendous affliction, to his acceptance of it as a compassionate act of love.

Despite the explicit and implicit messages, and prevailing social acceptance of mercy killing of people with disabilities, there is widespread rejection of the legitimacy of 'euthanasia' within the disabled community.[479] As the late Australian disability rights advocate Stella Young[480] argued, the advent of legalised euthanasia would create a false equivalence between a life without dignity and death without dignity, and grant a paternalistic medical profession the literal power of life and death over people whose lives appeared of a lesser value and apparent suffering.[481]

———

Throughout human history, a range of social, political and medical beliefs have underpinned the ways in which societies have demarcated 'disability' and differentiated people living with disability. From antiquity through to the Middle Ages, disability was seen as a form of divine punishment or demonic possession leading to persecution or confinement. By the time of the Enlightenment there was a shift in the demonising societal attitude towards people with disabilities, accompanied by the scientific study of, and provision of, institutionalised care for people with disabilities, including the introduction of 'moral treatments'. There followed in the eighteenth and nineteenth centuries the introduction of legal protections for people with disabilities, although many in society continued to objectify and ridicule the disabled, often exploiting them. In the first half of the twentieth century, the emergence of eugenics as a philosophy and practice, and its linking with racial science, saw the increasing segregation of people with disabilities including exclusion from the community and through restrictions in immigration.[482]

In Germany, societal approaches to people living with disability have evolved from sympathy and compassion for wounded soldiers from the 1914-1918 war in the Weimar years, through the virulent ableism of the Nazi period, to the gradual evolution of a disability rights movement, paralleling the social equality movement from the 1960s.[483] The late 1970s saw the emergence of disability activist groups, including so-called *Krüppelgruppen* (Cripples' Groups). The liberalisation of German federal politics after re-unification saw improvements in legislation that protected the human rights of people living with disabilities, including the imposition of quotas on companies and state institutions for employing people with disabilities. In 2017 the *Bundestag* legislated the *Bundesteilhabegesetz* ('Federal Participation Act') that protected equality in social participation and improved access to resources, financial support and peer support for people living with disabilities.[484] The German discourse of disability rights and the recognition of '*Ableismus*' (ableism) is rooted firmly in the legacy of the *Krankenmorde*, particularly German society's long-standing opposition to euthanasia, sterilisation and prenatal genetic testing.[485] The same historical legacy has, however, also led to many Germans living with disabilities to not disclose their problems to co-workers, employers or government agencies.[486]

The traditional definition of 'disability' follows a biomedical model, expressing it as a deficit affecting normative function in a person.[487]

This conceptualisation of disability posits that a pathological process leads to dysfunction of a body system or systems (disease), which impairs function (impairment) that translates into non-normative social functioning for the individual involved (disability). In this way a person born with cerebral palsy is considered to have suffered a brain injury or developmental anomaly during gestation or birth or the immediate period following. This leads to damage to the central nervous system which, depending on the region affected, translates into impairment of movement, speech, fine motor control and in some more extreme circumstances, cognitive functioning. If this person is unable to walk, manipulate objects or communicate using spontaneous or fluent speech, their 'disability' emerges during their encounters with others in daily life, particularly within the immediate community, and in important long-term areas of development and participation such as in education, interaction with peers, employment and career progress or other dealings with the social system.

The medical model of disability defines a person in terms of the functional limitations they have and the consequent social role into which they are cast. This approach to disability has cultivated the 'deficit' view that people living with disabilities have 'something wrong with them'.[488] Historically, this medically defined social role of incapability has led to a negative valuation of the personhood of people living with disability—recall the phrases 'life unworthy of life' or 'three generations of imbecile is enough'.[489] The devaluation of people living with disabilities is a significant factor in their significant social disadvantage.[490]

This devaluing perception has played a critical role in entrenching ableism. Unpacking the concept of ableism, Sandra Levi writes that while it is used to describe 'prejudicial attitudes and discriminatory behaviors toward persons with a disability', definitions of ableism 'hinge on one's understanding of normal ability and the rights and benefits afforded to persons deemed normal'.[491] Citing a fellow disability scholar, she writes that ableism 'has become a term used to describe "the set of assumptions and practices that promote unequal treatment of people because of apparent or assumed physical, mental, or behavioural differences"'. Levi considers ableism to be constituted by: lowered expectations in areas such as education or employment; 'normalisation as beneficence', such as promoting verbal over non-verbal communication by people with disabilities; limitations faced by people with disabilities in their self-determination; and the consequences of the label 'disabled'. Levi identifies contemporary applications of eugenics as including: preventing people with disabilities from reproducing or from being born; exclusion and isolation of people with disabilities through institutionalisation; and, in extreme circumstances, murder.[492] In its most gratuitous form ableism occurs through targeted violence (defined in some jurisdictions as a 'hate crime') or the use of pejorative language such as the epithets 'retard', 'moron' or 'spastic' to diminish, ridicule or humiliate a person living with a disability. The tendency to refer to a person by their disability—for example, 'a schizophrenic' or 'an autistic' instead of describing them, for example, as 'a person living with schizophrenia' or 'a person with autism'—is a common process of objectification of a person that is ableist.

Discriminations such as refusing to employ a person living with a disability or failing to provide appropriate access to transport or public buildings are 'hard' forms of ableist discrimination. 'Soft' forms of ableism also exist. Well-intentioned violation of interpersonal boundaries such as solicitousness, misplaced pity, demeaning and unwelcome efforts to support or assist a person with a disability, or a patronising or infantalising tone of speech, are soft forms of ableism. Ableism in health systems manifests in the failure to provide healthcare services that are appropriate to the needs of a person with a disability, as well as the automatic assumption of diminished autonomy or capacity or a lesser quality of life and, by extension, 'invisibility' in decision making about health care choices affecting the person.

Ableism is a recent conceptualisation of a phenomenon that was first recognised during the apogee of the eugenics movement. Randolph Bourne, an early twentieth century American writer who lived with severe physical disabilities, wrote in his 1911 essay *The Handicapped – By one of them*: 'The deformed man is always conscious that the world does not expect very much from him. And it takes him a long time to see in this a challenge instead of a firm pressing down to a low level of accomplishment. As a result, he does not expect very much of himself; he is timid in approaching people, and distrustful of his ability to persuade and convince. He becomes extraordinarily sensitive to other people's first impressions of him. Those who are to be his friends he knows instantly, and further acquaintance adds little to the intimacy and warm friendship that he at once feels for them. On the other hand, those who do not respond to him immediately cannot by any effort either on his part or theirs overcome that first alienation.'[493] The more recent introduction of the term 'ableism' evolved from the civil rights movements in the United States and Britain during the 1960s and 1970s, pushing prevailing perspectives of disability into a political paradigm.[494] This shift emphasised the role of society in the construction of discourses of and attitudes towards disability, as distinct from an individual bearing 'primary responsibility for enduring or remedying the disability through prayer in the religious paradigm or through medical intervention in the scientific paradigm'.[495] However prejudice and discrimination against people with disabilities is conceptualised, its ramifications continue to challenge progress in societies around the world.

More than one billion people live with psychiatric, physical and intellectual disabilities in the world today and most experience great social disadvantage.[496] A World Health Organisation (WHO) and World Bank report on disability observes: 'Many people with disabilities do not have equal access to health care, education, and employment opportunities, do not receive the disability-related services that they require, and experience exclusion from everyday life activities'.[497] The report conceptualises disability as an interaction between health conditions and the personal and environmental contexts. The experience of disability varies greatly and while disability correlates in general with disadvantage, not all people with disabilities are equally disadvantaged.[498] Women, children and adults with intellectual, sensory or mental health difficulties, and people with more severe impairments, experience greater disadvantages. The WHO-World Bank report identifies widespread evidence of barriers that restrict participation for people with disabilities. These include inadequate social policies, standards and funding in areas such as education, built environment, transport and communications; and employment or poverty reduction. They also face negative attitudes such as beliefs and prejudices among health care workers, teachers and employers; or family members that discriminate against, silence, or exclude people with disability. Most live in countries that lack provision of essential care services, or have poor and inadequate service delivery in areas such as health care, rehabilitation, or support and assistance. The report emphasises the lack of consultation with, and involvement of, people with disabilities in important decisions that affect their health and welfare. These barriers contribute overwhelmingly poor, and inevitably costly, health outcomes, 'including greater vulnerability to preventable secondary conditions and co-morbidities, untreated mental health conditions, poor oral health, higher rates of HIV infection, higher rates of obesity, and premature mortality'.[499]

These findings highlight the damage caused by ableism. People with disabilities generally have lower levels of educational attainment, participate less in the economy, experience higher rates of poverty and often cannot participate in activities in the general community. The personal and social consequences of intended and collateral ableism escalate: 'Reliance on institutional solutions, lack of community living, inaccessible transport and other public facilities, and negative attitudes leave people with disabilities dependent on others and isolated from mainstream social, cultural, and political opportunities'.[500]

The WHO-World Bank report proposes a model of disability framed as a 'workable compromise' between the medical and social models of disability. The report notes that '(d)isability is complex, dynamic, multidimensional, and contested', the authors suggesting that while the two dominant models of disability 'are often presented as dichotomous…disability should be viewed neither as purely medical nor as purely social: persons with disabilities can often experience problems arising from their health condition. A balanced approach is needed, giving appropriate weight to the different aspects of disability'.[501]

In 2006, the United Nations adopted the Convention on the Rights of Persons with Disabilities (UNCRPD). This has since been widely accepted as the definitive instrument to enable uniform international standards for human rights for people living with disabilities, including mental illness.[502] The convention 'finally empowered the world's largest minority to claim their rights, and to participate in international and national affairs on an equal basis with others who have achieved specific treaty recognition and protection'.[503] At the time of writing, the UNCRPD had been signed by 187 countries and ratified (made legally binding) by 177. By December 2018, an Optional Protocol to the UNCRPD that established a complaints mechanism had achieved 93 country signatories and 94 state parties to the protocol. The United Nations' Department of Economic and Social Affairs has summarised the UNCRPD, noting that:

'countries must guarantee that 'persons with disabilities enjoy their inherent right to life on an equal basis with others (Article 10), ensure the equal rights and advancement of women and girls with disabilities (Article 6) and protect children with disabilities (Article 7). Children with disabilities shall have equal rights, shall not be separated from their parents against their will, except when the authorities determine that this is in the child's best interests, and in no case shall be separated from their parents on the basis of a disability of either the child or the parents (Article 23). Countries are to recognize that all persons are equal before the law, to prohibit discrimination on the basis of disability and guarantee equal legal protection (Article 5). Countries are to ensure the equal right to own and inherit property, to control financial affairs and to have equal access to bank loans, credit and mortgages (Article 12). They are to ensure access to justice on an equal basis with others (Article 13), and make sure that persons with disabilities enjoy the right to liberty and security and are not deprived of their liberty unlawfully or arbitrarily (Article 14). Countries must protect the physical and mental integrity of persons with disabilities, just as for everyone else (Article 17), guarantee freedom from torture and from cruel, inhuman or degrading treatment or punishment, and prohibit medical or scientific experiments without the consent of the person concerned (Article 15). Laws and administrative measures must guarantee freedom from exploitation, violence and abuse. In case of abuse, States shall promote the recovery, rehabilitation and reintegration of the victim and investigate the abuse (Article 16). Persons with disabilities are not to be subjected to arbitrary or illegal interference with their privacy, family, home, correspondence

or communication. The privacy of their personal, health and rehabilitation information is to be protected like that of others (Article 22). On the fundamental issue of accessibility (Article 9), the Convention requires countries to identify and eliminate obstacles and barriers and ensure that persons with disabilities can access their environment, transportation, public facilities and services, and information and communications technologies'. [504]

The trajectory for this international acceptance can in large part be traced to determined agitation around disability in the last decades of the twentieth century. The social model of disability first emerged in the 1970s and 1980s when activists and scholars sought to redefine disability away from the traditional medical model. The Union of the Physically Impaired Against Segregation set out an influential account of the social causes of disability: 'In our view, it is society which disables physically impaired people. Disability is something imposed on top of our impairments, by the way we are unnecessarily isolated and excluded from full participation in society. Disabled people are therefore an oppressed group in society'.[505] Among those who continued to build on this work was British academic Mike Oliver, who argued that the notion of disability developed as the result of the interaction between people living with various impairments and an environment filled with physical, attitudinal, communication and social barriers. Oliver's 'social model of disability' proposes that rather than focusing on fruitless and frustrating attempts at rectifying the individual impairments that underlie disability, as we would with a disease, it might be better to work towards altering the social environment to enable people living with impairments to participate in society on an equal basis with others.[506] For Oliver, the phenomenon of 'disability' emerges from systemic perceptions of mobility and disadvantage that had become embedded in society rather than an inability or functional impairment of the individual.[507]

Despite the intuitive appeal of Oliver's thesis, the medical model persists, and by focusing on so-called 'deficits' and 'deviance' it continues to constrain the lives of people with disabilities.[508] Such ableist attitudes influence the approach of health professionals to people living with disabilities seeking health care.[509] Western culture categorises disability as a personal or family tragedy, punctuated by moments of 'heroic triumph'[510], when someone living with a disability performs successfully an otherwise mundane action such as maintaining a job or competing in a sporting event—becoming what some call 'inspiration porn'.[511]

The socio-political challenges facing people living with a physical, intellectual and psychiatric disability have arguably not retreated since Mike Oliver's vision of a 'social model' of disability was first elaborated.[512] Writing on the subject 30 years later he observed: 'The disabled peoples' movement that was once united around the barriers we had in common now faces deep divisions and has all but disappeared, leaving disabled people at the mercy of an ideologically driven government with no-one to defend us except the big charities who are driven by self-interest. Because of this, most of the political campaigning that has taken place in defence of our benefits and services have forced disabled people back into the role of tragic victims of our impairments and has involved others undertaking special pleading on our behalf'.[513] In his criticism Oliver was observing the malignant influence of market driven attitudes to social care that had developed under neoliberal policies which had gained favour in western economies around the time of Oliver's initial work.[514]

Neoliberalism, as a political and economic philosophy, relies on assumptions of personal responsibility and freedom to participate in consumer choice. Applied to health and disability services, neoliberalism recasts 'care' as one of many service commodities traded in an open market and, in the worst cases, such neoliberal governments resile from provision of services, leaving everything to market forces. Neoliberal influences on community relationships tend to separate people in economic roles of 'provider' or 'consumer' and measure human worth by its relative utility within global consumer culture. Alongside the neoliberal inspired reformulation of the relationship between disabled and non-disabled people was the dictum of 'no rights without responsibilities' and the expectation of participation in the economic system in exchange for social equity and equality.[515]

Neoliberal influences on disability rights included shifting the provision of disability support services from the state to the market; reframing disability through the creation of a class of 'able-disabled' capable of working in supported disability employment; and the marginalisation of people with disability who cannot work.[516] Therein lie similarities to the early twentieth century German asylum system following the introduction of '*arbeitstherapie*' and the dichotomous view of devaluing those people with disability who could not work or participate in any form of economically productive activity.

In 1922 English novelist GK Chesterton wrote of a future dystopia in his monograph *Eugenics and Other Evils: an argument against the scientifically organized state*. Chesterton predicted that future eugenic laws, such as those emerging in the USA and Europe at the time, would lead to widespread abuse and persecution of the poor and socially undesirable. The eugenic inspired abuses of human rights that followed were to prove Chesterton's anxieties correct.[517] The 1990s US film *Gattaca* depicts something akin to Chesterton's envisioned dystopia. The citizens of this eugenic society are divided into those pre-selected or engineered prior to birth for the best possible traits and those born through the lottery of 'natural' conception. The genetically inferior 'in-valids' face widespread discrimination perpetrated by the genetically elite 'valids'. The film's moral injunction is for us to define personhood beyond genetic makeup.

———

FIGURE 36 A disability rights poster commissioned by the World Health Organisation (WHO) This is one of seven posters that 'allude to the barriers which prevent people with disabilities from living their lives to the fullest. Each poster highlights one of the main areas for action covered in the World report on disability: health, rehabilitation, support and assistance, enabling environments, education and employment'.

Perhaps the most visible legacy of early twentieth century eugenics is the concept of 'reproductive rights' which, within the broader discourse of international human rights, has been framed as freedom of reproduction for all people. The World Health Organisation (WHO) states that 'reproductive rights rest on the recognition of the basic right of all couples and individuals to decide freely and responsibly the number, spacing and timing of their children and to have the information and means to do so, and the right to attain the highest standard of sexual and reproductive health. They also include the right of all to make decisions concerning reproduction free of discrimination, coercion and violence'.[518]

The original aspiration of twentieth century eugenics was to use scientific knowledge and technology to improve the human species over generations by altering, where necessary, its biological composition. Eugenicists sought 'to encourage people of good health to reproduce together to create good births (what is known as 'positive eugenics') and to end certain diseases and disabilities by discouraging or preventing others from reproducing ('negative eugenics')'.[519] The distinction between positive and negative eugenics dates to the turn-of-the-twentieth-century racist and eugenic writings of British physician Caleb Williams Saleeby (see Chapter 2).[520] Positive eugenics involved promoting the reproduction of desirable human germ lines primarily through means such as social engineering, where economic or other incentives were used to encourage reproduction of certain 'desirable' groups. Negative eugenics included measures such as the prohibition of marriage between those at risk of bearing genetically diseased children, enforced sterilisation or abortion and, ultimately, mass eradication of undesirable germ lines.

In the immediate post-war period, negative eugenics was categorised among the evils of the Nazi regime and yet involuntary sterilisation of people in prisons or psychiatric institutions continued.[521] Eugenics did not receive significant public attention again until later in the twentieth century when advances in reproductive technology enabled a new era of negative eugenics.

In contrast, positive eugenics lingered as a palatable option in the community in both private and public spheres. In the 1970s for example, Singaporean social and economic policies were enacted to increase the reproduction rates of educated and successful women and reduce that of low paid, uneducated women. In the United States in the late 1970s, the multimillionaire Robert K Graham established the 'Repository for Germinal Choice' in the Californian city of Escondido. Graham sought to establish a sperm bank replete with the germ of Nobel laureates and the highly intelligent, and later sought to broaden his criteria to include successful athletes.[522]

Negative eugenics reappeared in policy and public debate in various countries in the 1990s. In China, the *Maternal and Infant Health Law* (1995) made it illegal for people carrying heritable mental or physical disorders to marry, and promoted mass prenatal ultrasound testing for birth defects. At the same time in the United States, Richard Hernstein and Charles Murray published their polemic *The Bell Curve: Intelligence and Class Structure in American Life*, arguing that intelligence was strongly determined by genetics and race and had a proven predictive validity for success in life.[523] The book has many troubling negative eugenic social policy nostrums, such as measures to prevent women of low intelligence reproducing and to reduce immigration to the United States. When compared to the enforced sterilisation of the 'genetically inferior' in many countries and later mass murder of the disabled by the Nazis, the book's arguments appear to those so disposed as plausible social policy, with sentiments redolent of that of mid-twentieth century eugenics.

Among the more gratuitous manifestations of negative eugenics were those employed by the US charity 'Children Requiring a Caring Kommunity' (C.R.A.C.K.), which later reformed as the organisation 'Project Prevention'. Founded in 1997 by Californian woman Barbara Harris, who had adopted four children from the same drug-addicted mother, Harris sought unsuccessfully to have the Californian law changed to criminalise drug use by pregnant women. The main focus of Project Prevention has been to provide access and financial reward for drug addicted men and women to undertake sterilisation or long-term birth control. In support of her charity, Harris infamously said: 'We don't allow dogs to breed. We neuter them. We try to keep them from having unwanted puppies, and yet these women are literally having litters of children'.[524]

―――

The birth of Louise Brown on 25 July 1978 in Manchester heralded a new era in eugenics. Louise was the first child born following the in vitro fertilization (IVF) of her mother, a technique pioneered by two fertility experts from Cambridge University, Robert Edwards and Patrick Steptoe. The success of this form of assisted fertility saw its widespread adoption to the point where, if the technology is available and affordable, it is almost an automatic choice for couples struggling to start a family. The advent of Gamete In Vitro Fertilization (GIFT) and Zygote In Vitro Fertilization (ZIFT) has enabled many older women to carry pregnancies to term. However, despite the technical advances in assisted fertility, it remains costly, emotionally and physically draining and not highly successful. In some jurisdictions this has meant that restrictions have been placed on access to assisted fertility, creating another category of bioethical dilemma around equity and equality.

Another access point for the re-emergence of negative eugenics was the completion in 2006 of the seven-year international scientific collaboration to map the molecular sequence of human DNA, known as 'The Human Genome Project'.[525] This has enabled biotechnical companies to appropriate genetic codes for certain desirable and undesirable human traits. There are now commercially viable clinical applications for prenatal identification of genetic disease and disorders. Given the high stakes in assisted fertility, preimplantation genetic diagnosis (PGD) and screening (PGS) have become the norm in many countries.[526] In some instances PGD is being pursued as a critical part of public policy: in China, for example, the government has made reproductive medicine, including PGD, a spending priority[527] to improve the genetic health of the nation. There appears a fundamental difference between current practices of PGD and the eugenic excesses of the first half of the twentieth century, distinguished around the availability of individual reproductive choice. In the example of China's funding priority for PGD there is a determined application of state control and here we can see biopower and governmentality at work: in the state policies and the social and economic incentives that embed positive and negative eugenics as a preferred control over the population.

The utilisation of prenatal genetic testing has been variously described as 'family eugenics', 'private eugenics' or 'new-genics'.[528] The term 'newgenics'—as a signifier for the use of such testing and the choice to terminate genetically problematic pregnancies—is itself controversial. While from one perspective 'newgenics' may represent a form of negative eugenics, others may see it as exercising a right to a pregnancy outcome that is for the betterment of the family and the community. How one chooses to frame prenatal genetic testing as either adding an ostensibly healthy person to the community or sparing the community the cost of a 'genetically compromised' person underlines the complex social and moral terrain facing newgenic use: where does the distinction lie here between positive or negative eugenics? Adam Cohen, whose book on American eugenics has influenced this work, has weighed in on this debate arguing that, 'Twentieth century eugenics has rightly been called a 'war on the weak'—its goal was to stop people with conditions like Huntington's disease from reproducing. Twenty-first century eugenics can enable people with the Huntington's gene to have children without it. The new eugenics can be a war for the weak'.[529]

In her research into genetic counselling offered by professionals in the late twentieth century, academic Dorothy C Wertz concluded that 'eugenic thinking survives'.[530] Though the usual practice in genetic counselling is to provide the patient with non-directive and unbiased information to help autonomous decisions, Wertz found that the norm among clinical geneticists in 36 countries was to provide 'directive pessimism', including urging patients to terminate some genetically diseased pregnancies, or presenting them with selective information. Her body of work also highlighted the inconsistent approach of geneticists to defining the 'severity' of the around 600 identified genetic disorders, determinations critical in patients' decisions to end or continue pregnancies.

After the first use of PGD in 1989, the term 'reprogenetics' appeared in the field. The advent of PGD was 'the use of genetic technologies in the context of reproduction to select what kind of children to bring into the world'.[531] Medical geneticists seek to be non-directive in light of a 'history of gruesome eugenics'. Yet PGD and newer reprogenetic practices such as non-invasive prenatal testing (NIPT) for screening and CRISPR genome editing technologies for direct intervention on embryos seem to rekindle fear of a resurgent eugenics within both society and the academy. This fear raises the question of whether labeling a practice 'eugenic' warrants its moral condemnation.[532]

While genome editing is not currently utilised for human reproductive purposes[533] NIPT has been steadily integrated into prenatal care in nearly 90 countries.[534] NIPT uses a maternal blood sample to test for chromosomal conditions such as Trisomy 21 (Down syndrome), Trisomy 18 (Edwards syndrome) and Trisomy 13 (Patau syndrome). NIPT is more accurate than previous techniques of amniocentesis or chorionic villus sampling. The degree of utilisation of NIPT in different countries is as much a function of cultural and religious views as of cost.[535] The commercialisation and commodification of NIPT has, however, been the main factor influencing its introduction into markets in the US, China/Hong Kong, Western Europe, the Middle East, South America, Asia and Africa.[536] NIPT is projected to be a US $2.5 billion industry by 2025[537], with much of this growth to be realised in a Chinese market defined by its 'two-child policy'.[538] With NIPT promising such an enormous return on investment, the biotechnology industry—not the medical profession—has become the main driver in its use in antenatal care.[539] In the example of NIPT we can observe both the critical importance of control over reproduction[540] and the increasing influence of neoliberalism and market economics in the present day exercise of biopower[541].

In 2016, the UK's National Health Service (NHS) announced the roll out of NIPT. This soon generated concern about the potential misuse of NIPT to identify the gender of the fetus— an approach that could easily lead to the morally questionable practice of sex selection as sought by some parents.[542] In response to the NHS announcement, British actor Sally Phillips produced the documentary *A world without Down's syndrome?* for the BBC. Phillips' had multiple concerns and motivation: given that around 90 per cent of families who receive a prenatal diagnosis of Down syndrome choose not to progress the pregnancy, the introduction of NIPT could lead to the eradication of Down syndrome from the human species.

This was a deeply personal issue for Phillips, as one of her sons, Olly, lives with Down syndrome. She compares him to the Dodo bird, 'curious, friendly, gentle', and likens the medical geneticists keen to eradicate Down syndrome as the explorers who bludgeoned the bird into extinction. In a poignant scene in the documentary, Phillips interviews Halldora, an Icelandic woman living with Down syndrome. Halldora speaks of the apparent diminution in the value of her life, especially as 100 per cent of families in Iceland now choose to terminate fetuses identified as having Down syndrome. This new reality, as the Icelandic geneticist Kári Stefánsson details later in the film, is 'merciless'. As Phillips poses: if a fetus with Down syndrome is so devalued by Icelandic society, what does this say about Halldora's life and citizenship in the eyes of her fellow citizens? Disability scholar Leah Burch argues that the film touches on important and 'previously silenced' issues: the fostering of a societal narrative that Down syndrome is tragic and disastrous; increasing societal pressures to abort fetuses with Down syndrome; the (in)adequacy of public debate regarding the ethical issues of prenatal screening; the exclusion from such debate of the voice and experience of individuals living with Down syndrome; and whether new, more accurate prenatal screening will affect 'who we want to live in this world?'.[543]

Bioethical analysis of reprogenetics highlights the complexity and nuance within debates about NIPT. The expectation among some bioethicists is that introducing NIPT will modify rather than 'revolutionise' PGD.[544] The prospect of a more accessible process of genetic manipulation of the species frames the bioethical debate around NIPT as incorporating reproductive rights, disability rights and the historical precedent of twentieth century— eugenics. The at times polarising debate over whether testing in pregnancy properly accommodates disability rights 'tends to manifest as an *impasse* between disability advocates and test advocates. Disability advocates voice concerns about the disappearance of the conditions screened for. Test advocates have a tendency to frame disability medically, using terms such as 'suffering' and 'difficulties".[545]

The apparent moral and ethical permissibility of a reproductive technology appears to reflect how 'eugenic' it may seem.[546] One of the problems situating the questions of reproductive rights within the framework of eugenics is that the latter, as typically understood, is applied at the level of the population rather than discretely as a specific reproductive choice within an individual family. This distinction prompted American molecular biologist Joshua Lederberg to argue in the 1960s for the use of the term 'euphenics' (literally, good or normal appearance) to apply to the process of improving the features of individuals and families through genetic manipulation.[547] This has manifest in the advent of so-called 'designer babies', facilitated by the potential for molecular biotechnology to enable a future pregnancy to be possessed of desirable traits. While this has already emerged in the controversial phenomenon of sex selection applied within some cultures, the prospect of selecting desirable genetic traits in a future person poses as many bioethical questions as negative eugenics. This strays into the concept of 'transhumanism' popularised by the biologist Julian Huxley.[548] Transhumanism is, at its essence, the improvement of human potential through enhancement, whether it be genetic, technologically assisted, or through physical modification. Julian Huxley's account of transhumanism—a concept that can be traced back to antiquity—is disturbingly redolent of the dystopia described in his brother Aldous Huxley's 1932 novel *Brave New World*, and later appropriation by the Nazis of Nietzsche's 'Übermensch' (Super human)[549] within their eugenic and racist project. Transhumanism exceeds eugenics in its scope, prompting bioethical debates over moral, cognitive, perceptual or physical enhancements, both before and after birth.

The present state of molecular genetics and the practical application of this to human reproduction embodies the kind of modernity Zygmunt Bauman described—the attempted exertion of human control over nature.[550] The suppressed premise in these types of seemingly transhumanist reproductive choices, whether population or individually based, is the relative valuation of life they may reflect. This returns us to the fundamental question raised both by eugenics and the *Krankenmorde*: is there to be a relative valuation of kinds of life and by extension, kinds of persons?

―――

The reciprocal influences of eugenics and modern genetics are explored in a memorial exhibition at the site of *Am Spiegelgrund* in Vienna, one of the deadliest *Kinderfachabteilung* in the *Krankenmorde*.[551] The contemporary exhibition housed at the site alerts visitors to the dangers of reductionist biological explanations of human life as offered currently by medical science:

'These new trends promise to supply biological explanations for mental phenomena that seemed unthinkable twenty years ago. "Biological psychiatry" seemed obsolete then - today it seems to represent a utopian hope for cures and possibly even prevention. As yet, however, we cannot assess whether these findings will live up to their promise. The utopia of "healing" has already once ended in the gruesome reality of "annihilation." On a totally different level, "racial hygiene" (or "eugenics") has paved the way for today's human genetics. While we now recognize that National Socialist crimes were based on a wholly inadequate understanding of heredity and its highly complex mechanisms, modern genetics promises to supply this certain knowledge "at last." The utopia of a "genetically healthy" humankind seems once more within reach.

However, the information promulgated in simplified form amongst the public often does not correspond with the complex findings of science. Our old images of "biology" and "heredity" threaten to direct the new knowledge along the old lines. Were the Nazis right after all? Are we returning to the "biological psychiatry" of the Nazi period? The diverse results of modern research do not indicate that and reveal a much more complex reality. But they are interpreted within the old frames of biologistic thinking. This is where a critical approach and democratic public debate are urgently needed.'[552]

Lisa Hempel, Emil B, Sonia Wechlser, Werner Przadaka, Georg Mall and the hundreds of thousands of others who perished in the *Krankenmorde* were murdered because their lives were—within a rapid period of time and through secretive policy escalation—deemed inferior; reframed as a genetic and economic problem and publicly expressed through negative symbolism and propaganda. Their lives were terminated brutally via the available industrial technologies—for the 'good' of themselves, their family and the newly conceived 'pure' Germanic nation. Similar crimes could yet recur as nations seek to impose their preferred population characteristics and hierarchies both within and outside their borders. As we have seen, the paradigm continues—there is a deep compulsion to control the natural world and bioshape humans within a particular society, especially to create and populate 'healthy and desirable' families and nations based on binary biological conceptions of 'health' and 'illness' or a 'good' or 'bad' life. These aims remain a prime driver in most nation's social health policies.

Under the *Krankenmorde* those condemned as 'feebleminded' and 'mentally retarded' and other medical categories regarded appropriate at the time, died under the malign influence of phenomena that we now see as 'ableism' and 'biopower'—rationales and practices that become embedded in controlling institutions of state and continue to inform the development of bioscience, genetics and population control. The motivating forces behind the philosophies of 'eugenics' and 'euthanasia' are evident in the gathering influence of molecular genetics applied in public policy and increasingly, by parental and individual choice.

If born in present times, most of these people would still face lives of exclusion, social disadvantage and a shorter life expectancy than their fellow citizens. Their biology might well be controlled through contraception or sterilization, psychotropic medication and coercive psychiatric treatment. Their fellow citizens might accept the legitimacy of a 'good death' in preference to a life of lesser value, accepting the reasoning and associated risks that these assisted deaths are legitimate clinical interventions. Yet this would be—and continues to be—the defining existential question in their/our lives.

Given the cost obsessed flow of social policy, of institutional acceptance and promulgation of preferred population characteristics, and the economic challenges in supporting people who fall outside a dominant 'normal' spectrum, they would continue to be defined by their incapacities and experience social disadvantage, reduced longevity and second class citizenship. And all this would occur in a liberal democracy that espouses individual freedom and human rights, and at the same time asserts categorically that it abhors the policies and practices employed by the Nazi state. To acknowledge this is to begin to apprehend the contemporary ethical significance of the *Krankenmorde*.

CHAPTER 9
REMEMBRANCE, COMMEMORATION, MEMORIALISATION

Babette Fröwis was born in Munich in July 1929. Her father, Wilhelm, worked as a labourer on a dairy farm. From birth, Babette demonstrated numerous feeding and settling problems and spent the first five months of her life in an institution for children with disabilities. Babette returned to live with her family and continued to show signs of significant developmental delay—she did not walk until the age of three and by age ten she had never spoken, other than to babble like an infant. She was not able to learn to use the toilet. As a child she suffered numerous seizures and exhibited increasingly aggressive behaviour, including tearing out her hair and screaming uncontrollably. Babette's parents became concerned that her behaviour posed a risk to her younger siblings. Paediatricians declared her an 'imbecile' and 'ineducable'. In August 1934 Babette was placed in permanent institutional care at the Schönbrunn Sanatorium in the city of Dachau. This 100-year-old institution was then under the Caritas umbrella, operated and managed by the Munich Catholic diocese. Babette remained at Schönbrunn until late 1943. In early October the medical director of Schönbrunn, a paediatrician named Dr Hans-Joachim Sewering, informed Babette's parents that due to her behaviour she could no longer be properly cared for at Schönbrunn. On 23 October 1943 Sewering signed a transfer order for Babette to be sent to the *Kinderfachabteilung* at the hospital in Eglfing-Haar on the outskirts of Munich.

Babette Fröwis died there on 16 November 1943. In her Eglfing-Haar medical file an entry reads 'inadequate food intake for five days, frequently chokes while eating. In the last few days tracheobronchitis. Died today'.[553] Despite this statement, Babette had been assessed as being of robust physical health when admitted to Eglfing-Haar three weeks earlier. The lies documented in her medical file were intended to conceal the fact that she had died after being overdosed fatally on a medication, most likely Luminal.

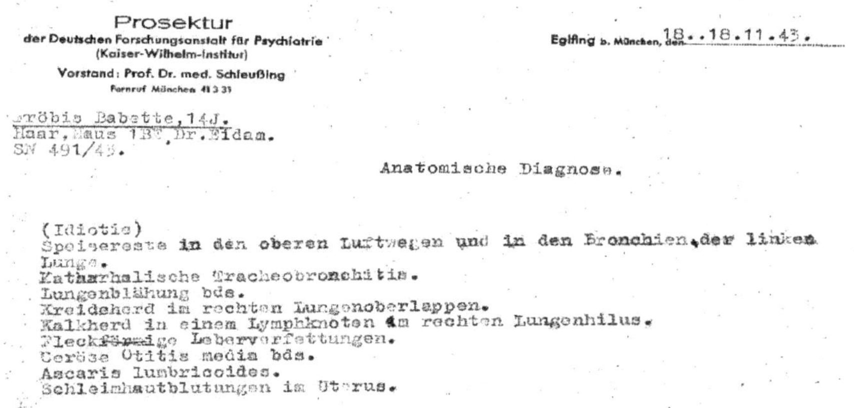

Prosektur
der Deutschen Forschungsanstalt für Psychiatrie
(Kaiser-Wilhelm-Institut)
Vorstand: Prof. Dr. med. Schleußing
Fernruf München 41 3 31

Eglfing b. München, den 18..18.11.43.

Fröbis Babette, 14J.
Haar, Haus 1B , Dr. Eidam.
SN 491/43.

Anatomische Diagnose.

(Idiotie)
Speisereste in den oberen Luftwegen und in den Bronchien, der linken Lunge.
Katharrhalische Tracheobronchitis.
Lungenblähung bds.
Kreidherd im rechten Lungenoberlappen.
Kalkherd in einem Lymphknoten am rechten Lungenhilus.
Fleckförmige Leberverfettungen.
Seröse Otitis media bds.
Ascaris lumbricoides.
Schleimhautblutungen im Uterus.

J.V.

FIGURE 37 The post mortem report for Babette Fröwis dated November 1943

The post mortem report stated that she died of a lung infection in Eglfing-Haar.

Hans-Joachim Sewering was born in Bochum in 1916 to a working-class family. He gained entry to medical school in Vienna and graduated in 1941. He worked as a staff paediatrician at Schönbrunn and was later appointed to the position of medical director of the institution. Sewering had joined the Nazi Party as a medical student in 1933 and later joined the SS.

After the war, he presented himself to a denazification court which determined him to be a middle level functionary in the regime. Sewering received his denazification certificate after paying a modest fine. In 1947 he returned to his paediatric practice in Dachau and joined the Christian Social Union, an influential Bavarian political party active in medical politics. It is unclear to what degree Sewering's medical and political colleagues in the period turned a blind eye to his wartime activities. In 1955 he assumed the leadership of the Bavarian Medical Association and in 1968 was appointed as a professor of paediatrics in Munich. Sewering's career in German medical politics reached its peak in 1973 when he became President of the national *Bundesärztekammer* (BÄK or German Medical Association), the successor association of the Reich Physicians' Chamber established under the Nazi regime's *Gleichschaltung*. At the time, the post of a BÄK president also included the responsibility to serve as Treasurer of the World Medical Association (WMA).

Over time it became evident that Sewering had participated in the children's 'euthanasia' program. In 1978 the magazine *Der Spiegel* published a series of articles on the topic of the children's 'euthanasia program' that linked the death of Babette Fröwis to Sewering. While resigning his position at the BÄK that year,

Sewering continued to deny knowledge of what had happened to the children he had authorised be sent to Eglfing-Haar and other such institutions. Sewering's actions under the Nazi regime were again mentioned in 1989 in a publication dealing with Nazi medicine.[554] In 1993, four nuns who had worked previously with Sewering accused him of signing the transfer orders of more than 900 children to euthanasia institutions such Eglfing-Haar[555].

Despite the controversy around Sewering, the WMA rewarded his long service in medical politics by electing him president at its October 1992 meeting in the Spanish resort town of Marbella. The WMA had been formed in Paris in 1947 primarily in response to the disturbing revelations of medical crimes from the Nuremberg Doctors' trial. The WMA had formulated a 'physician's oath' in Geneva the following year that, among other things, proclaimed that physicians ought to respect all human life. Sewering's presidency of the WMA caused a stir, particularly among Jewish physicians in the organisation. Members of the WMA such as Michael Kochan from Göttingen, William Seidelman from Canada, and Michael Grodin from the United States, agitated for Sewering's removal from the position of President-Elect of the WMA.

Sewering stood his ground, giving a somewhat self-serving interview to the influential newspaper *Süddeutsche Zeitung* in January 1993, claiming that both he and the nuns at Schönbrunn knew nothing of the fate of the children sent to Egfling-Haar.[556] Sewering's interview prompted the Catholic Bishop of Munich to issue a statement which indicated that staff at Schönbrunn knew that the approximately 900 children sent to Eglfing-Haar and similar institutions would be killed. As a result, Sewering had little option than to resign his position at the WMA, although he framed his act as one of self-sacrifice to protect both the WMA and BÄK against 'threats from the Jewish World Congress'.[557] Sewering's defenders even cited Babette Fröwis' allegedly problematic behaviour in an attempt to justify his actions.[558] In 1994 the US Department of Justice determined Sewering's status as a war criminal and banned him from entering the United States.

Despite the international condemnation for his wartime conduct, Sewering's stocks remained high among his German colleagues. In January 1996 the journal *Deutsches Ärzteblatt* published a notice to honour the occasion of Sewering's 80th birthday. This prompted a blistering rejoinder from William Seidelman in the April edition of the same publication.[559] In 2008 Sewering was honoured for his services to the German medical system.[560] After his death in 2010, Sewering's obituary in the *Deutsches Ärzteblatt* referred to his 'leadership in medical ethics'[561].

In 2011 Frank-Ulrich Montgomery, a radiologist working in Hamburg, assumed the Presidency of the BÄK. A year later, the BÄK proclaimed its 'Declaration of Nuremberg'.[562] The document sought to emphasise the moral responsibility of German physicians in the Nazi period, stating: 'in contrast to assumptions still widely held today, the impetus for these most grievous violations of human rights did not originate from the political authorities, but from physicians themselves. These crimes were not the acts of individual physicians, but were carried out with the participation of leading representatives of the established medical profession and professional medical societies, as well as with significant involvement by distinguished medical academics and members of renowned biomedical research institutions'. While the BÄK placed responsibility for these medical crimes with both the medical profession and the individual physicians, critics of the 2012 Nuremberg declaration consider it incomplete because it omits to acknowledge the German medical profession's journey from compulsory sterilisation to the extermination camps in Poland.[563]

To date, the World Medical Association remains silent on the Sewering affair[564].

The BÄK's 2012 Declaration of Nuremberg was made at a time when another significant cultural movement was occurring in a re-unified Germany, part of a larger social and cultural process of engagement with the Nazi period.

Questions of German guilt and responsibility for the crimes of the Nazi regime evolved over several post-war generations. The psychoanalyst Alexander Mitscherlich formulated an influential account of the phenomenon of German guilt for the Nazi years. After he had observed the Nuremberg Doctors' Trial (1946-47), Mitscherlich later wrote about Germany's 'inability to mourn' its Nazi past.[565] Mitscherlich argued that, in the immediate post-war years and then the subsequent period following the break-up of their country in 1948, Germans lacked empathy for the victims of the Nazi regime. While in the 1950s citizens of the German Democratic Republic (GDR) came to terms with their situation behind the 'iron curtain', West Germans were engrossed in the rapid economic recovery (the *Wirtschaftswunder*) of their new liberal democracy. In such circumstances Germans had little time or motivation to engage with the victims of their Nazi past. Mitscherlich proposed that Germans considered the Nazi years as something that happened *to* them and that to permit any psychological internalisation of either collective guilt for crimes like Treblinka or Hadamar, or empathy for the millions of victims of Nazism, would be profoundly destructive to the collective national psyche, hence the 'inability to mourn'.

The victorious Western Allies had, in the first instance, implemented an inchoate and desultory process of 'denazification' of the millions of Germans who had been members of Nazi organizations living in the western occupied zones. The exigencies of the Cold War caused the process to peter out in the 1950s. From 1949 to 1963 the series of governments in West Germany led by Chancellor Konrad Adenauer of the Christian Democratic Union initiated a new cultural and political process known as *Vergangenheitsbewältigung*—'coping with the past'. Through school curricula, social institutions and publicly funded arts and cultural institutions, West Germany under Adenauer made attempts to engage with its guilt and responsibility for the crimes of National Socialism.[566] Not all were persuaded of the wisdom or rectitude of Adenauer's cultural project. The members of the neo-Marxist 'Frankfurt School' of social and critical theory were unconvinced, with one of the most prominent members, Theodor Adorno, deeply sceptical about *Vergangenheitsbewältigung*. In November 1959, Adorno gave a speech in Wiesbaden following the 'Swastika epidemic'—a series of anti-Semitic attacks by Neo-Nazi groups that disturbed West Germany's sensibilities.[567] Adorno argued that the process of working through the past was deeply flawed and that 'the past that one would like to evade is still very much alive. National Socialism lives on, and even today we still do not know whether it is merely the ghost of what was so monstrous that it lingers on after its own death, or whether it has not yet died at all, whether the willingness to commit the unspeakable survives in people as well as in the conditions that enclose them'.[568]

The notion of the distinctness of 'German guilt' has seen an equally controversial debate. German historians such as Fritz Fischer[569] and Hans-Ulrich Wehler[570] argued in the 1980s that the German militarism that led to the 1914-1918 war and the socio-cultural conditions that led to the Nazi period were the result of a specific form of German 'modernity'. Germany's development in the nineteenth century saw the evolution of a form of economic liberalisation without full political and social liberalism. This theory is generally termed the *Sonderweg* or 'special path'. While Germany's economy industrialised and developed, there remained in German society a residual *Klassenhabitus* (class awareness) manifesting as a servile attitude to the persisting social power of the aristocratic class. The failure of German liberals to capitalize on the wave of revolutions against the ruling elite that occurred in Europe in 1848 was, to paraphrase British historian AJP Taylor, the historical turning point where Germany failed to turn. Taylor went on to argue that the *Sonderweg* is most evident in the observation that the only government created by modern Germans for themselves was the Nazi regime.[571]

In the 1980s the issue of German guilt for the Nazi period was the focus of a culture war between the German political left and right wings, played out through the media, academia and social commentariat in what was referred to in Germany as the *Historikerstreit* (historians' quarrel). The right-wing view of the Nazi period, championed by historian Ernst Nolte, was that Germany's actions in the war were no worse than the USSR's and that the Nazi's war against Stalin was one of national survival. The left-wing response was typified by Frankfurt School academic Jürgen Habermas' view that Nolte's analysis was an attempted whitewash or 'cancelling out of damages'[572].

The specifically German origin of the Holocaust is an equally divisive issue. Many European countries had long traditions of anti-Semitism and colluded with Germany in perpetrating the attempted genocide of their Jewish populations. The question of a specific form of German anti-Semitism as both a necessary and sufficient preconditions to the Holocaust is, however, a controversial and polemic debate. The issue came under intense consideration in the wake of the publication in 1996 of American writer Daniel Goldhagen's book *Hitler's Willing Executioners*.[573] Goldhagen argued that there was a distinct and virulent form of German anti-Semitism that extended in a historical process dating from Martin Luther, reaching its apogee in Hitler's regime. Goldhagen argued that most Germans at the time enthusiastically prosecuted a genocidal war of elimination, particularly in the east.

Recognition of the suffering of Europe's Jewish population under the Nazi regime entered collective awareness in the international community at the time of the trial of war criminal Adolf Eichmann in Israel in 1961.[574] Subsequent popular interest in the period emerged during the Frankfurt Auschwitz trial (1963-65) in which more than 200 survivors provided compelling testimony of the crimes perpetrated in the Holocaust. In 1966 Albert Speer, Hitler's former minister for armaments and war materiel, was released from prison and spent the rest of his days building his celebrity as 'the good Nazi' by giving interviews and publishing his reflections of the Nazi period.[575] It was during this time that a renewed interest within Germany in Adolf Hitler as a historical figure developed.[576]

A key moment of widespread German engagement with the history of the Nazi period arose in January 1979 when the national West German broadcaster ZDF aired the US-produced mini-series *Holocaust* (1978), featuring the then-emerging Hollywood stars Meryl Streep and James Woods.[577] While ZDF had previously aired several locally produced documentaries about the Nazi period, the network's decision to purchase for $US600,000 the foreign rights for the production from the American NBC network was politically controversial. The broadcast of *Holocaust* brought about 'an unlikely national catharsis' in Germany and it precipitated extensive discussion of the subject in the public sphere.[578]

In the 1990s a new generation of Germans was confronted by its Nazi past when the Hamburg Institute for Social Research launched a travelling exhibition 'War of Annihilation. Crimes of the Wehrmacht 1941 to 1944'.[579] The 'Wehrmacht Exhibition' opened in Hamburg in March 1995 and travelled to 33 German and Austrian cities. More than 800,000 people visited the exhibition. The Wehrmacht's alleged participation in the Holocaust had been a topic of academic research from the late 1970s[580], however the Wehrmacht Exhibition was the first public assertion that the regular German army was complicit in genocidal violence on the Eastern Front, and that the memory of the humble soldier or *Landser*[581] as merely defending his homeland was false; that he was not as distinct from the 'evil SS' as had been depicted in years past. This revived on a much broader social canvas the historical debates of the 1980s, confirming the view that the war in the East was one of racial elimination in which the entire military establishment was implicated.[582] The initial Wehrmacht Exhibition was halted when it became clear there were some historical inaccuracies in the content and a revised version was launched in Berlin in November 2001.

The Nazi period has become a mainstream cultural focus in Germany, part of a phenomenon now termed as *Erinnerungskultur* (the culture of remembering). This is evident in film, television, literature, scholarship and art that has emerged in recent years, and also in the dedicated focus given in numerous museums, statues, memorials and art work found in Berlin and other major German cities.[583] In more recent *Erinnerungskultur*, the Nazi period has been addressed in several celebrated German film and television productions, such as *Downfall* (2004), *Sophie Scholl - The Final Days* (2005), *Generation War* (2013) and *Labyrinth of Lies* (2014)[584].

The *Krankenmorde* has, by contrast, been under-represented in popular German culture. A minor subplot of the 1979 *Holocaust* TV series featured the murder of the character of Anna Weiss at the Hadamar 'euthanasia' centre after she suffered a nervous breakdown following a sexual assault. In literature, Elisabeth Claasen's memoir, *Ich, die Steri* (I am the Sterilised), discusses her sterilisation under the hereditary health laws. Alois Kaufmann's novel *Totenwagen* (Death Wagon) depicts the experience of a child living with disabilities in a reform school and refers to the Am Speigelgrund *Kinderfachabteilung* in Vienna. Elvira Hempel's self-published memoir *Die Hempelsche* (The Hempel Girl) tells of her escape from the Brandenburg gas chamber and the difficult years that followed, including her troubled mid-life.[585] Robert Dome's 2008 novel *Nebel im August* (Fog in August) provides a narrative with many parallels to the story of Elvira Hempel.[586] Dome's protagonist, Ernst Lossa, is born to a Yenish family (German Roma). Ernst's mother dies in pregnancy and his father is imprisoned in a concentration camp as the family is considered socially undesirable. Ernst is placed into an orphanage where he is stigmatised as racially undesirable and a maladjusted criminal. He is later sent to the Kaufbeuren-Irsee hospital as part of the 'decentralised euthanasia' program where he is murdered. The book was made into a feature film released in 2016.

―――

The management of memorials, sites of remembrance, and museums that function as institutions of memory of the Holocaust and the Nazi period, has been particularly challenging. In despairing of the apparent 'kitsch' of present-day tourism to Auschwitz, Dutch-American writer Ian Buruma cites the German concept *entweihen* (profanity), which he considers to mean 'to rob something of its sacred nature'.[587] Many Holocaust survivors or their descendants are often offended or appalled at the behaviour of visitors to memorial sites in Germany and Poland—the regular scenes of indifferent young people smoking, drinking or petting on top of the stelae of Berlin's 'Memorial to the Murdered Jews of Europe' (Holocaust Memorial)[588] or the infamous 2014 'Auschwitz selfie' of an Alabama teenager seem to prove Buruma's point.[589] Many such sites of 'death tourism' have succumbed to a process of 'Disney-fication', whereby the visitor is entertained as a 'consumer'.[590] More still, such sites can become politicised or focal points for protest or rallying of neo-Nazi or anti-Semitic groups. In locations outside of Europe, Holocaust memorials and museums function as 'secular sacred spaces': sites of mourning, remembrance, education and community engagement for Holocaust survivors and their families forced to leave Europe[591].

Against these challenges, Germans have attempted to represent and commemorate the victims of the *Krankenmorde* in different ways, particularly since reunification. German memorials fall into different categories. A *Mahnmal* (memorial) such as Berlin's Holocaust Memorial, serve the purpose as a general warning from the past, whereas a *Denkmal* (monument) serves as a reminder of historical events—a specific form of *Denkmal* is a *Gedenkstätte* (commemorative site) which is located on the actual site of an historical event or incident.

Most German cities have some form of memorial to the victims of the Nazi period and increasingly to the victims of the T4 program. In December 1992 a Berlin artist, Gunter Demnig, laid the first of thousands of *Stolpersteine* (stumble stones), part of a project that places cobble-stone sized concrete cubes with commemorative brass plaques at the last known address of a person who perished under the Nazi regime. By the end of 2016 there were more than 60,000 of the privately-sponsored *Stolpersteine* laid at 1,800 sites in Europe, and more than 15 *Stolperschwelle* (larger 'stumbling thresholds') for certain cases where hundreds or thousands of individual *Stolpersteine* would have to be laid in a single place. Four *Stolperschwhelle* were dedicated to T4 victims, while the number of *Stolpersteine* dedicated to individual *Krankenmorde* victims could not be confirmed.[592]

In 2005 two artists, Horst Hoheisel and Andreas Knitz, designed a commemorative sculpture they named *Denkmal der Grauen Busse*— The Grey Bus Memorial. The memorial consists of two concrete buses split along their lengths to symbolise the *Gekrat* buses used to transport patients to T4 killing centres. On the inside of each memorial is an inscription that questions:

'*Wohin bringt Ihr uns?*' (Where do you take us to?)

One of the two concrete buses remains as a permanent installation at a psychiatric clinic in Weißenau, whilst the second is moved and installed in different locations across Germany.[593]

The site of the former Liebermann Villa at *Tiergartenstraße* 4 is now in front of the *Berliner Philharmonie*. The original villa was destroyed and the Berlin Philharmonic Hall was built on the site in the 1960s. After much delay and community advocacy, the *Bundestag* decided in late 2011 to construct a permanent memorial to the victims of 'Nazi euthanasia' at the *Tiergartenstraße* 4 site. The T4 *Gedenkstätte*—Memorial for the Victims of the Nazi 'Euthanasia' Program—was officially dedicated in September 2014.

At the sites of all six of the T4 killing centres there are *Gedenkstätten*. On the site of the Brandenburg killing centre, in place of the prison complex which was destroyed during the war, a new building containing a multimedia installation provides visitors with a detailed account of the T4 program and related history in the region. The area surrounding the memorial installation marks the site of the former gas chamber, undressing rooms and crematorium. The *Gedenkstätten* at Hadamar and Bernburg retain intact the original gas chambers and surrounding rooms. There are *Gedenkstätten* at the killing centres at Pirna-Sonnenstein and in the castle at Hartheim where the original building structures also remain intact. At both sites visitors are guided through museum installations that focus primarily on the victims. The Hartheim memorial site also houses a permanent display on the topic of human rights. The site of the Grafeneck killing centre was returned to the Samaritan organisation after the war and is now a supported residential community for adults living with disabilities.

FIGURE 38 **The Memorial for victims of the Nazi 'Euthanasia' Program on the original site of the *Tiergartenstraße* 4 villa, Berlin**
The Berlin Philharmonie building is at the rear of the structure.

Many hospital sites in Germany, Austria and Western Poland now have memorials and museums dedicated to the patients who perished in the *Krankenmorde*[594].

FIGURE 39 Trees on the hillside below the Pirna-Sonnenstein memorial site
Trees have been painted to show the area where the cremains of victims of *Aktion* T4 and *Aktion* 14f13 were deposited from the crematorium.

The inevitable question that emerges with any process of national atonement for state-sponsored crimes is that of compensation for the victims. As Elvira Hempel was to learn in her later years, the compensation process for survivors of the *Krankenmorde* and the families of the victims was both prolonged and complicated.[595]

Survivors of compulsory sterilisation and the *Krankenmorde* faced several legal, cultural and political barriers to both acknowledgement of and compensation for their suffering.

Post-war juridical processes in West Germany against perpetrators of the *Krankenmorde* were intermittent and tended to be overshadowed by more publicised trials of war criminals involved in the mass murder of Jewish and other racial victims in death camps. The gradual progress of engagement with the *Krankenmorde* accelerated after German reunification allowed greater access for scholars and researchers to archival holdings in the former Communist bloc.[596]

The crime of 'euthanasia' was part of the original indictment in the Nuremberg 'Doctors' trial' and three defendants—Karl Brandt, Viktor Brack and Waldemar Hoven[597]—were executed after their convictions, which included involvement in the *Krankenmorde*. Other smaller scale trials of *Krankenmorde* perpetrators followed. In October 1945 the United States attempted the prosecution in Frankfurt of several medical and nursing staff from the Hadamar killing centre. Prosecutors in the Hadamar trial anticipated the defence lawyers' argument that the killing of these patients was lawful as it occurred under the 1934 German Hereditary Health Laws. This compelled the prosecution lawyers to focus on the murders of nearly 500 *Ostarbeiteren* (foreign workers) at the Hadamar site as part of *Aktion 14f13* and not seek convictions for any of the murders of *Krankenmorde* victims. A subsequent series of trials of alleged *Krankenmorde* perpetrators were held in Frankfurt between 1946 and 1948.[598] In various West German states, courts heard cases against medical and nursing defendants on the murder of more than 15,000 Germans killed as part of the *Aktion* T4 or decentralised 'euthanasia' programs. The prosecution cases avoided the supposed legality of actions under Hereditary Health Laws and concentrated on whether the accused's actions constituted a violation of natural law and moral conscience. There were no attempts in West Germany or the GDR to prosecute those involved in compulsory sterilisation.

In 1964, Fritz Bauer[599], the state prosecutor from the state of Hesse, prepared a detailed indictment against the chief T4 psychiatrist Werner Heyde. Heyde had escaped custody after the war and practiced for 14 years under the alias 'Dr Sawade' in the state of Schleswig Holstein in northern Germany. Many of Heyde's colleagues knew of his crimes and covered for him. Heyde often gave expert testimony in disability benefit claims, including those of victims of the Nazi regime. After his identity became known (in part due to the involvement of Hans Creutzfeldt—famous for Jacob-Creutzfeldt disease), Heyde surrendered to police. He underwent many hours of interrogation by Bauer and his colleagues. The document that emerged from this interrogation, the 'Bauer Indictment' is now an invaluable source of information about *Aktion* T4. He hanged himself in February 1964, five days before his trial was to start.[600]

In 1955 the Allied powers imposed a 'Transitional Agreement' on the new Federal German Republic (West Germany) as an interim compensation scheme. Part Four of this agreement obliged West Germany to pay compensation to victims of the Nazi regime who were persecuted based on race, politics or religion. Victims of the hereditary health laws who were compulsorily sterilised were not part of this arrangement. In subsequent years the stigma of mental illness within the community often prevented families of victims speaking out or seeking compensation. Many psychiatrists who had been perpetrators of crimes committed in the *Krankenmorde* had returned to clinical practice after the war and, disturbingly, a number sat on tribunals that heard claims for compensation for the very crimes they had been involved in committing. Many claimants were dismissed by these tribunals, having been deemed unreliable witnesses because of their mental health problems[601].

The process of compensation for victims of the hereditary health laws and the *Krankenmorde* evolved in three stages over the seventy years after 1945. In the 1950s and 1960s, *Bundes-Entschädigungs-Gesetz* - BEG (Federal Act for Compensation of Victims of National Socialism) established that any damage suffered by a claimant had to be directly linked to the 1934 hereditary health law. In the 1970s and 1980s a series of Social Democratic governments in Germany sought to correct this limited view of history, provoking a wider conversation in German society about the 'forgotten victims' of National Socialism. Even then the *Bundestag* did not wish to recognise any obligation to pay financial compensation to victims of the Hereditary Health Laws and agreed to a one-off DM 5000 payment to claimants to put an end to the matter.

From the 1990s onwards another generation of Germans, the grandchildren of the Nazi period, re-engaged in the significance of that time and the implications for German society. In 1998 the 'Law for Abrogation of National Socialist Wrongful Judgements in Criminal Justice' quashed criminal convictions for *Widerstand* (resistance) martyrs, such as Dietrich Bonhoeffer and the Scholl siblings.[602] It also overturned all judgements made by the Hereditary Health Courts established after the 1934 *Law for Prevention of Hereditary Diseased Offspring*. It was not until the occasion of the United Nations International Holocaust Remembrance Day on 22 January 2011 that the *Bundestag* more than doubled benefits for victims of compulsory sterilisation. In January 2017 the *Bundestag* focused on the victims of the *Krankenmorde* in its commemoration of that year's Holocaust Remembrance Day.[603]

———

One of Frank Schneider's clearest memories of his childhood in the 1960s was the abject state of the adults and children living with disabilities in the State Hospital at Weilmünster near Frankfurt-am-Main in the German state of Hesse. Schneider was the son of a building contractor and grew up in the nearby town of Wetzlar. He recalled often looking through a large window in a building in the Weilmünster hospital and seeing that the patients with physical and intellectual disabilities were neglected, living in overcrowded wards and excluded from the rest of the community. These memories would become a profound influence on Schneider's later professional life.

Frank Schneider's family had not lost a relative in the *Krankenmorde*. His mother, an ethnic German from Katowice in Poland, fled the advancing Red Army with her mother and sister in 1945, one of the millions of *Flüchtlinge* (refugees) that poured into Germany at the end of the war as part of a forced mass migration now termed *die Vertreiben* (the eviction). His maternal grandfather had been taken prisoner by the Soviets and sent to Siberia until his return to Germany in the late 1940s.

Growing up, Schneider had joined the *Jugendrotkreuz* (German Youth Red Cross) and entered his medical studies in the 1980s intending to become a paediatric neurologist. Instead, his career took him into psychiatric training and his later professional achievements in the field of neuroimaging culminated in his appointment to a professorship in the University of Aachen in western Germany.

In 2009 Schneider was elected to the Presidency of the *Deutsche Gesellschaft für Psychiatrie und Psychotherapie, Psychosomatik und Nervenheilkunde* (German Society for Psychiatry, Psychotherapy, Psychosomatics and Neurology or 'DGPPN'). The DGPPN had its origins in a preliminary professional organisation formed by German psychiatrists in 1846 in Kiel, becoming the *Verein der Deutschen Irrenärzte* (Society of German Doctors for the Insane) in 1864. In 1903 the Society renamed its discipline 'Psychiatry'. The profession would subsequently integrate with neurologists during the Nazi period. After German re-unification, the professional societies of psychiatry and neurology in the former East and West Germany came together as the DGPPN.[604]

At the time of Schneider's election to the presidency of the DGPPN, the German medical profession had little expectation that medical students and doctors in training would, as a matter of course, be provided with education about its history. After the Sewering scandal of the 1990s a new generation of German physicians, particularly paediatricians, began to engage with their history in the Nazi period. One of Schneider's main priorities as President of the DGPPN was to draw upon the experiences of his youth to bring his profession with him in engaging with its past, not only to atone for the crimes perpetrated by psychiatrists in the Nazi years, but to also to document and historicise them. Schneider sought to embed within the profession both education and research about the lessons of this historical period of psychiatry and work towards a formal apology for the crimes of psychiatrists in the Nazi era.

At the time of Schneider's election, 15 former presidents of the DGPPN were still involved in the organisation. Many had direct links with the Nazi period and believed the matter closed. The stigma faced by the German psychiatric profession, already considerable because of its wartime history, had intensified, given the ongoing controversy about its powers of coercion in German society and the ability to force involuntary hospitalisation upon people living with mental illness.

FIGURE 40 Frank Schneider delivers the formal apology of the German Psychiatric Association to victims of the *Krankenmorde* and their families in Berlin on 26 November 2010

Schneider believed that as an important first step, the DGPPN needed to alter its constitution to acknowledge its history and encourage research and education about the Nazi period. Schneider's strategy in negotiating with his senior colleagues was to make clear that the process of an apology by the DGPPN and attempts at reconciliation would occur eventually, and he encouraged his senior colleagues to be a part of it. Following a period of discussion and persuasion, Schneider prevailed on both the question of an official apology being offered by the DGPPN and modifying the organisation's constitution to acknowledge the historical significance of the Nazi period. The Executive Board endorsed the DGPPN position on 23 November 2010 and Schneider delivered the apology in Berlin three days later. In an oration that narrated the history of the crimes of psychiatrists under National Socialism, Schneider stated on behalf of his colleagues:

'In the name of the German Association for Psychiatry and Psychotherapy, I ask you, the victims and relatives of the victims, for forgiveness for the pain and injustice you suffered in the name of German psychiatry and at the hands of German psychiatrists under National Socialism, and for the silence, trivialisation and denial that for far too long characterised psychiatry in post-war Germany'.[605]

To commemorate the occasion, Dorothea Buck, an artist who had been a victim of compulsory sterilisation under the 1933 Hereditary Health Law, gifted Schneider a copy of her famous statue *Mutter und Kind* (Mother and Child).[606]

The DGPPN had also resolved that it should not write its own history and following the 2010 apology it consulted with historians from the Topography of Terror organisation and the Foundation Memorial to the Murdered Jews of Europe. This productive collaboration generated numerous memoria including the travelling poster exhibition 'registered, persecuted, annihilated – the sick and disabled under National Socialism' launched at the *Bundestag* in 2014. This German-English-language exhibition was specifically aimed at a wide audience and posed deep questions on:

the value of life as a guiding principle; the complex interactions that contributed to the era; the intellectual and institutional prerequisites for the murders; the events of exclusion and forced sterilisations up to and including mass extermination. It deals exemplarily with victims, perpetrators, persons involved and opponents, and reflects on the analyses of those events from 1945 to the present. The exhibition, which has been shown around Germany and internationally, succeeds in providing victims and their families with a voice and a testament of their suffering and death.

The arrival of the DGPPN travelling exhibition at different sites often stimulates local, academic and community events that focus on the topic of the persecution and murder of the sick and the disabled. These enable discussion about the contemporary challenges facing people living with disabilities. At one such event in Aachen in the summer of 2016, Frank Schneider visited a school that provides specialised educational programs for children living with various forms of disability. When the children performed a concert for the community he was profoundly moved, recalling his own childhood memories of the abandoned souls in the Weilmünster institution. Here, instead, the children were singing 'I am different, you are different, we are different', celebrating the fact that their individual challenges, personal triumphs, joys and frustrations are a critical part of the human spirit.

EPILOGUE

After escaping death at the Brandenburg killing centre in September 1940, Elvira Hempel spent six months in the paediatric psychiatry unit at the Görden institution. In March 1941 she was returned to *Haus* 50 at Uchtspringe and after a few months there, was moved to *Haus* 49, where it seemed to her that the residents were less disabled than those in *Haus* 50. Being relocated within Uchtspringe would be another unsettling stroke of good fortune as, by the end of the war, a further 750 children from *Haus* 50 were murdered by starvation or lethal injection.[607] Elvira remained ever hopeful that Lisa would join her there, but when her mother made one of her rare visits to Uchtspringe in September 1941, Elvira learned that Lisa had died. In fact, Lisa was murdered in the Brandenburg killing centre's gas chamber in August 1940 and like so many other families of the victims of *Aktion* T4, it is probable that Elvira-Lotte had received a letter stating a false cause of death.[608] Despite a brief period of leave at home, Elvira's mother was unsuccessful in persuading the authorities to release Elvira permanently.

For the next few years, Elvira was moved between several different welfare homes and allowed to attend a few years of school. In 1942 she was forced to join the girl's equivalent of the Hitler Youth, the *Bund Deutscher Mädel* (League of German Girls). In 1943 she returned to live with her mother again and, within the limitations of their difficult relationship—particularly Elvira's resentment of a much younger sister Kaale who, as the apple of her mother's eye, was enjoying a privileged childhood by contrast to Elvira's—she was able to experience some independence and a period of 'normal' childhood. This was again to be short lived as conditions deteriorated rapidly within Germany during 1944, especially in the vicinity of Berlin.

After the war ended, Elvira reunited with her father Otto who, despite abandoning Elvira-Lotte and his family, invited Elvira to join him in Berlin with his new partner and child. This proved to be little more than a ruse for Elvira to nurse the child while they pursued work in the destroyed capital. When Otto's new partner turned against Elvira, she was abandoned again, and without the critical ration card. Alone in the Soviet occupied part of Berlin, Elvira found spartan lodgings at a convent and for the next few years was unable to sustain any meaningful employment.

In 1946 Elvira's mother left Magdeburg for Berlin where she and Elvira later reconciled. They lived together in the cellar of a bombed-out building with meagre food and clothes and desperate to avoid the occupying Soviet forces. Elvira's mother helped her find employment dismantling factory equipment that was sent back to the USSR. They lived in constant fear of sexual assault by Red Army troops and this often led to Elvira missing work. Of the nearly two million German women who were raped by Red Army troops in the latter stages of the war and during the occupation, around 240,000 of these victims were attacked in Berlin. Around 10 per cent of the women raped by Red Army soldiers died as a result of this mass-scale sexual violence.[609] Unable to feel safe or find stable work in Berlin, Elvira and her mother returned to Magdeburg in 1947.

Elvira found it difficult to form personal or intimate relationships with men during this period of her life. However, in 1950, she met Rainer Schultz through mutual friends and they started a romantic relationship. Having learnt of the mass sterilisation of the 'feebleminded', Elvira believed that she too had been sterilised by exposure to X-rays and was unable to fall pregnant. In this she was mistaken, and in October 1950 she gave birth to a daughter, Angelika. Schultz soon abandoned them and Elvira was left distraught—weakened by a severe post-partum infection and left to raise Angelika alone, she contacted a social worker for support. Without knowledge of the trauma of Elvira's first decade of life, the social worker recommended that Elvira either give Angelika up for adoption or place her in an institution. This triggered an unfettered sense of rage in Elvira, who accused the social worker and her profession of complicity in her sister's death. The trauma of motherhood for Elvira was at its most salient when Angelika demonstrated gross motor developmental delay and plagiocephaly[610] at 12 months of age. Through her own child she was being forced to not just relive her own childhood but also to recall the many children in *Haus* 50, including her sister Lisa, who had similar problems. However, in confronting these fears, Elvira was able to recognize a deep love and affection for her suffering daughter. She was determined to persevere.

In 1952 Elvira met Harry Baum and after a brief romance they married. She obtained a driver's permit and found work as a bakery delivery driver. The rushed marriage to Baum proved a disaster, however. Elvira would later describe an emotionally and physically abusive relationship with Baum, who she said often threatened to kill Angelika, referring to her as the *Huren Kind* (whore's child). Determined to protect Angelika and herself, Elvira fought back in what became an intolerable domestic violence situation. After surviving a near fatal overdose of barbiturates following a particularly ferocious encounter with Baum, Elvira and Angelika fled the household in 1955, finding shelter with one of her sisters in Flensburg, a small city in northern Germany.

Heinz Manthey was born in 1929 in the town of Stettin in Pomerania (present day Szczecin on the Polish-German border).[611] There his father managed a large estate and, after what he recalls as a happy childhood despite the wartime conditions, Heinz decided to become a teacher in 1944. Under the Nazi regime, school students who had completed eight years of school could streamline early into teacher training. Heinz commenced his studies in pedagogy in Köslin (Koszalin in Poland's North West). His teacher training ended abruptly in 1945 when he and his fellow students were conscripted into the *Volkssturm*, the militia or home guard comprising teenagers and elderly men, drafted by the Nazis to defend the Fatherland. Heinz's corps saw limited combat and he was captured by the US Army but later escaped, being lightly wounded in the process. After the war, he was unable to return to Soviet-occupied Pomerania and worked in a series of temporary jobs before becoming head of a department in a mechanical engineering company that repaired military vehicles in Flensburg.

In 1965 Elvira was employed as a process worker in the same company, leading to their meeting and starting a relationship later that year. Heinz recalls feeling 'butterflies' when he first met Elvira. They were engaged on *Silvester* (New Year's Eve) 1965 and, after Elvira's divorce to Harry Baum had been finalised, they married the following September. Having received a small inheritance, Heinz and Elvira purchased a house in the outer suburbs of nearby Lübeck. They later bought a fast food van and ran it as the family's primary income source. At other times, to supplement their income, Elvira drove a taxi in Lübeck and nearby Hamburg.

FIGURE 41 Hans and Elvira Manthey on their wedding day in September 1966

Heinz's family warmed to Elvira and Angelika immediately and by all accounts the marriage was happy throughout the next two decades. Elvira revealed little of her experiences during the war, saying only that she was 'in prison' as a child, although Heinz always doubted the story. In later years Elvira gradually revealed her childhood experiences. Heinz recalled her being tormented periodically by a nightmare where she was trapped in an air raid shelter or a prison cell or a gas chamber with two other girls. Her nightmares frequently featured the iron door to the Brandenburg killing centre's gas chamber. At other times, without provocation, Elvira would shut down, retreating into an intense melancholia. Heinz observed that she coped better when busy with work or family, only decompensating when left alone with her memories.

After they sold their fast food van business in 1984, Elvira's mental health deteriorated and, in the midst of a severe depressive illness, she made a suicide attempt. She was hospitalised in the psychiatric ward of the University clinic in Lübeck for several weeks and did not engage with further psychiatric treatment after she was discharged. Her condition remained fragile and, in 1988, Heinz recalled Elvira becoming 'frozen' on the lounge in their small living room as they watched a TV documentary about the Nazi 'euthanasia' program. She had recognised a photograph of the Brandenburg an der Havel prison complex during the broadcast and this in turn triggered powerfully dire memories and unresolved grief. After this she made another suicide attempt and was hospitalised, but again refused any follow up care after being discharged.

Despite the intense trauma of the realisation of her brush with death and the true fate of Lisa and the other children at Brandenburg killing centre, Elvira was emboldened to seek further information. In 1988 Uchtspringe and Brandenburg an der Havel were behind the 'iron curtain' in the German Democratic Republic (GDR), although this did not deter her from writing about her situation to the office of the state president of the GDR, Erich

Honecker. To Elvira's surprise, Honecker's office replied to her letter, providing details of the institutions she had been sent to, including her transportation to the Brandenburg killing centre. A year later Elvira applied successfully for a pension as a victim of the Nazi regime under the *Bundes-Entschädigungs-Gesetz* BEG (discussed in the previous chapter).

Following the collapse of the GDR in 1989 and the reunification of Germany, Elvira was able to visit the hospital at Uchtspringe in the summer of 1990 where, to her shock, she found her bed was still in its original place in *Haus* 50. The only acknowledgement of what had happened to all the victims at the institution was a small commemorative plaque describing the deaths of patients as *Gnadentot* (a mercy death). The hospital's director agreed that Elvira be given a copy of her medical file when she returned a few months later. Elvira and Heinz also drove to Brandenburg an der Havel only to find that the prison and the killing centre had been destroyed in the later years of the war. There was nothing to indicate what had occurred at the site.

Determined that the unnecessary and harrowing experiences of Lisa, herself and others be acknowledged by the government and community, Elvira resolved to complete a memoir, a project made nearly impossible by her limited education. After an abortive attempt to have the account ghost written by a professional author, Heinz took on the role of scribe, recording Elvira's narration of her early life. The self-published book, *Die Hempelsche*[612] (The Hempel Girl) appeared in 1994. As a consequence of her research, Elvira became concerned that her childhood diagnosis as 'feebleminded' placed her daughter and grandchildren at risk, should any future government seek to persecute people with hereditary illnesses or disabilities. Elvira and Heinz entered into a collaboration with an academic social worker researching the Nazi-era hereditary health laws, aiming to petition the Government to have Elvira's medical record amended. After a prolonged campaign for recognition and to have Fünfgeld's 1938 diagnosis of 'feeblemindedness' rescinded, the German Federal Petitions Committee heard Elvira's case in June 1996 and quashed the 'feebleminded' status.

Fifty-five years after the parents of Gerhard Kretschmer petitioned Adolf Hitler to have their 'disabled' son killed, Elvira Hempel-Manthey was released from her Nazi era medical sentence.

FIGURE 42 Elvira Manthey (front) visits the Uchtspringe Institution in the early 1990s

In the later years of her life, Elvira engaged with various political and religious groups for her emotional succour, and particularly where there were shared agendas to narrate the story of Nazi era medical crimes. The Manthey's home in Lübeck was opposite a Mormon Church. On several occasions during heavy snow falls in the late 1980s and early 1990s, the church caretaker also cleared a path through the snow to the front of the Manthey's house. In 1994 Elvira gifted the caretaker a copy of *Die Hemplesche* in gratitude for his kindness. The *bonhomie* between the Manthey's and the caretaker prompted the leaders of the Lübeck Mormon Church to invite them to join their congregation. Despite her long held anger at the Catholic Church from her childhood experiences in their institutions, Elvira attended the Lübeck Mormon Church congregation and seemed to find comfort in its community. Over the years the Mantheys would host American Mormon missionaries visiting Lübeck.

In the 1990s the Hamburg-based German chapter of the Church of Scientology took an interest in Elvira's story and her mission to make the public aware of the Nazi 'euthanasia' program. Her bitter experience in the hands of the established churches and the medical profession, especially psychiatry, were compelling, and provided a useful narrative for the expanding Scientology movement. The enmity of Scientology towards the psychiatric profession was laid down in the establishment of the 'Church of Scientology' in 1954 when founder L. Ron Hubbard published his theory of 'Dianetics', a potential rival treatment to the practice of psychotherapy. In effect, Scientology and psychiatry were competing for the same population of distressed and maladjusted people[613] and Hubbard would later proclaim that Scientology and psychiatry were at war—leading to the formation of initiatives against psychiatry, including the creation of the Citizens Commission of Human Rights (CCHR) in 1969.[614] In 1995 two German Scientologists, Thomas Röder and Volker Kubillus, published a book *Psychiatrists: The Men Behind Hitler*[615] arguing that the psychiatric profession had conspired to create the Holocaust as a first step to a program of world domination. In 2005 the Scientology-run CCHR established a museum on Sunset Boulevard in Los Angeles called 'Psychiatry: An Industry of Death'. Not surprisingly, the crimes of the psychiatric profession under the Nazi regime featured prominently in the CCHR's message.

It is little wonder that the publication of *Die Hempelsche* in 1994, and the shadow cast over the psychiatric profession, made Elvira a potential asset in Scientology's long standing anti-psychiatry agenda. Elvira was recruited as a speaker at several Scientology events, both in Europe and the United States. Her story helped make credible the movement's argument that psychiatrists were in large measure responsible for the Holocaust. However, being happily situated within the Mormon congregation in Lübeck, Elvira was not motivated toward a deeper engagement with the Scientology movement and her relationship with the organisation dwindled.

―――

Given her remarkable story, several historians also collaborated with Elvira. The then Oxford-based historian Michael Burleigh arranged for her to appear in his 1991 documentary film *Selling Murder: The Killing Films of the Third Reich*.[616] Elvira also collaborated on several projects about children's 'euthanasia' with a Frankfurt based writer Ernst Klee, a prominent researcher of the Nazi period. Sometime in the 1990s the Mantheys visited the 'Euthanasia memorial' in Bernburg. As Bernburg was one of the two former T4 killing sites where the gas chamber complex was still intact, it had a particular emotional salience for Elvira. There she met Ute Hoffmann, the chief historian at the memorial site, and later volunteered to participate in the education program at the centre. Elvira allowed some extracts from *Die Hempelsche* to be used for teaching purposes and she often spoke to Hoffmann's students. The collaboration worked well initially although, according to Hoffman, it faltered when Elvira insisted that the importance of her story was that the Nazis also tried to kill people like her, people who weren't disabled. Hoffman was concerned that this would legitimate the Nazi regime's programs in murdering people with disabilities. Hoffmann also believed that Elvira had been duped by both Scientology and some ultra-right wing political groups in East Germany who had provided her a platform to tell her story. The relationship between Elvira and Hoffman broke down after Hoffman believed that Elvira had made politically controversial statements to students and in public.[617]

———

Elvira's health deteriorated in 2010 after multiple heart attacks complicated her recovery from a series of hip operations. She suffered several strokes and developed dementia over the next 12 months. As the situation became impossible for Heinz to manage, in 2011 she was placed in an aged care facility in Lübeck. Elvira died in a Lübeck nursing home in 2014, aged 83.

When Heinz Manthey agreed to be interviewed for this book in the late autumn of 2016, he still lived in the Lübeck home he had shared with Elvira for decades. The house is filled with pictures of family, memorabilia, files and many copies of *Die Hempelsche* that he still sells on request. Herr Manthey impressed as a stoic man, deeply committed to his late wife's legacy. He emphasised that despite the many challenges he and Elvira faced, he believed that they had a good marriage and he had no regrets about the sacrifices he had made to be with her. Despite his own health problems, Herr Manthey had remained an active man, tending his garden and making his own wine and pickles. Elvira's grandchildren are regular visitors.

Like many of that generation of Germans, the lives of Elvira and Heinz Manthey were a mixture of trauma, privation, disruption, loss and shame, beneath which lay the same joys, tragedies, frustrations and mundanity experienced by others. Elvira Hempel's early years were spent in the most extraordinary and historically significant of circumstances—on the margins of the *Krankenmorde*. As its most recognised survivor, her life story is a public document of the Nazi persecution of the disabled. When we situate the intensely disturbing account of Elvira Hempel's childhood within this grand narrative, the uniqueness of her personal journey risks being obscured—as does the experience of any victim of the *Krankenmorde*. The details of Elvira's first years in Magdeburg and her gross social deprivation and illnesses and lack of a formal education, as well as the repeated betrayals, personal failings and criminality of many of those around her, all had a part in determining the personal difficulties she faced later in life—even allowing for the buffer of a loving and stable marriage to a decent man. It would add little to indulge in speculative 'psychohistory', to try and retrospectively diagnose any number of psychiatric disorders Elvira may have suffered later in life. To categorise her complex journey in such a way would seem to perpetrate the same psychiatric abuses as Dr Fünfgeld or the T4 assessors who condemned Elvira and Lisa Hempel to death.

Who is anyone to judge a child, or any person, and determine that they cannot live a worthwhile life? Are we not compelled by this story to ask what can we do to help all children realise their full potential?

Part 2 - Lessons and Legacies

FIGURE 43 Elvira Manthey visits the memorial on the site of the Brandenburg killing centre (1992)

The First Into The Dark

We dedicate this book to the hundreds of thousands who suffered or were murdered in the *Krankenmorde*. They too are the 'sentries' whose legacy stands guard to warn when cultures, societies, institutions or professions cross into the dark of *Dog Fox Field*.

ENDNOTES

A note on citation of references
A significant amount of the material cited is from the German scientific and humanities literature. The custom in such publications is to cite the initial of the first name of the author. To standardise the citations in this book, we have used the same convention for references from the English scientific and humanities literature in addition to any non-fiction or electronic texts.

CHAPTER I
THE IRON DOOR

1 K Synder (2001), 'Die Landesheilanstalt Uchtspringe und ihre Verstrickung in nationalsozialistische Verbrechen' ['The State Hospital Uchtspringe and their involvement in National Socialist crimes'], *Psychiatrie des Todes: NS-Zwangssterilisation und „Euthanasie' im Freistaat Anhalt und in der Provinz Sachsen (Teil 1)*, W Welz (ed), Magdeburg: Landeszentrale für politische Bildung des Landes Sachsen-Anhalt. W Ernst and T Mueller (2010), *Transnational Psychiatries: Social and Cultural Histories of Psychiatry in Comparative Perspective c. 1800-2000*, Cambridge Scholars Publishing.

2 E Manthey (1995), *Die Hempelsche - Das Schicksal eines deutschen Kindes, das 1940 vor der Gaskammer umkehren durfte*, Lübeck: Hempel-Verlag Heinz Manthey.

3 E Manthey (1995), ibid.

4   E Manthey (1995), op cit.

5   A Ley and A Hinz-Wessels (2012), *The 'Euthanasia Institution' of Brandenburg an der Havel*, Berlin: Metropol.

6   E Manthey, ibid; Elvira Manthey, *Testimony of Elvira Manthey - the Russell Tribunal on Human Rights in Psychiatry*: http://www.freedom-of-thought.de/rt/manthey.htm (30 June 2001) [Accessed 3 September 2016]; United States Holocaust Memorial Museum, *Oral History - Benno Müller-Hill, Antje Kosemund, Paul Eggert, and Elvira Manthey*, USHMM: https://www.ushmm.org/wlc/en/media_oi.php?ModuleId=0&MediaId=5090;

7   E Manthey (1995), op cit.

8   E Manthey (1995), (2001), op cit.

9   S Knittel (2016), 'Autobiography, Moral Witnessing and the Disturbing Memory of Nazi Euthanasia', *Reverberations of Nazi Violence in Germany and Beyond*, ed. M Fulbrook, S Wienand, and J Christiane Wagner, pp.65–82

10 M Burleigh (2000), *The Third Reich - a New History*, London: MacMillan.

11 D Vyleta (2011), *The Quiet Twin*, London: Bloomsbury.

12 M Burleigh (2000), ibid.

13 E Manthey (1995), op cit.

14 Dr Prof E W Fünfgeld, *Report On Patient Elvira Hempel September 9 1938 – Official Copy of Medical Record*, Landesheilanstalt Uchtspringe. Courtesy of Heinz Manthey.

15 E Manthey (1995), ibid.

16 S Knittel (2016), ibid.

17 Records show that on the day, 52 women and 22 children and youth under 18 years of age were deported from Uchtspringe to Brandenburg. See K Synder (2001), ibid.

18 The term '*Tötungsanstalten*' (killing centres) is now used for these sites, although at the time they were referred to as 'Institution A, B', etc.

19 The term 'Nazi' is used as an abbreviation of 'National Socialist' from the *Nationalsozialistische Deutsche Arbeiterpartei* (NSDAP).

20 From this point, we will refer to the killing centre at Brandenburg an der Havel as the 'Brandenburg killing centre' (or Tötungsanstalt), distinct from any references to the

town or the German state of Brandenburg and other public institutions sharing the name.

21   A Ley (2012), ibid.

22   A Ley (2012), op cit; H. Friedlander (1995), *The Origins of Nazi Genocide: From Euthanasia to the Final Solution*, Chapel Hill: University of North Carolina Press.

23   E Klee (1985), *Dokumente Zur Euthanasie*, Frankfurt-am-Main: Fischer Taschenbuch Verlag. The actual statistics about victims of the 'euthanasia program' can be found in the archive so-called 'Hartheim Statistics' in the National Archives and Records Administration (NARA) RG338, Microfilm Publication T-1021, Roll 18.

24   A Ley (2012), op cit.

25   F Schneider (2010), *Psychiatrie Im Nationalsozialismus*, Berlin: Springer. Within the German Reich it was estimated that about 70,000 people were killed in *Aktion* T4, 5,000 children murdered in *Kindereuthanasie*, 10,000–20,000 killed in *Aktion* 14f13, and 87,000 people also systematically murdered during 'de-centralised euthanasia'. In eastern occupied territories, the estimated 80,000–100,000 people murdered by the SS, SD and Wehrmacht units included patients killed in shootings and (mobile van) gassings, as well as those who were victims of starvation or similar deliberate fatal neglect and abuse. Variations in numbers can be partly attributed to different availability and interpretation of data, including in the categorisation of people as victims in specific actions. The extent of killing of institutionalised patients in other occupied parts of Europe (such as the Netherlands and France) is less well understood, with emerging research suggesting tens of thousands of people died from starvation and neglect. The following works contain more information and further references: Deutsche Gesellschaft für Psychiatrie und Psychotherapie, Psychosomatik und Nervenheilkunde (2014), *registered, persecuted, annihilated: The Sick and the Disabled under National Socialism - Exhibition Catalogue*, Berlin: Springer; and G Hohendorf (2016), 'The extermination of mentally ill and handicapped people under National Socialist rule', *Online Encyclopedia of Mass Violence*, [published online 17 November 2016], https://www.sciencespo.fr/mass-violence-war-massacre-resistance/en/document/extermination-mentally-ill-and-handicapped-people-under-national-socialist-rule

26   H Friedlander (1994), 'Step by Step: The Expansion of Murder, 1939–1941', *German Studies Review* 17 (3): pp.495–507

27   This statement needs to be qualified in that more than 90 Jewish victims died in the 'November Pogroms' (also known as 'Kristallnacht'). See M Gilbert (2006), *Kristallnacht: Prelude to Destruction*, New York: Harper Collins.

28 *The War* (2007), Directed and Produced by Ken Burns and Lynn Novik, PBS WETA, Washington, DC and American Lives II Film Project, LLC. http://www.pbs.org/thewar/Episode 7

CHAPTER 2
'LIFE UNWORTHY OF LIVING'

29 Adolf Hitler, 'Order to Bouhler and Dr. Karl Brandt to Increase the Authority of Physicians to Perform Euthanasia (Doc. PS-630),' Nuremberg Military Tribunal, NMT 01. Medical Case - USA v. Karl Brandt, et al.

30 Heeres-Sanitaetsinspektion im Reichskriegsministeriums (1934), *Sanitaetsbericht über das deutsche Heer, (deutsches Feld- und Besatzungsheer), im Weltkriege 1914–1918.* Volume 3, Sec 1. Berlin, pp.12–14

31 Keynes believed the punitive harshness of the Versailles treaty would have dire consequences and described it as a 'Carthaginian Peace' in reference to the sack of Carthage by Rome after its defeat in the Third Punic War in 146 BC, including the infamous account of ploughing the fields of Carthage with salt. See J M Keynes, *The Economic Consequences of the Peace* (London: Macmillan, 1920).

32 J Keynes (1926), 'The end of laissez-faire', *Essays in Persuasion*, London: Palgrave Macmillan, pp.272–294

33 The origin of the 'stab in the back myth' (*Dolchstoßlegende*) is attributed to the post-war remarks of Field Marshall Ludendorff. See J Wheeler-Bennett (1938), 'Ludendorff: The Soldier and the Politician', *Virginia Quarterly Review*, 14: pp.187–202

34 K Binding and A Hoche (1920), *Die Freigabe Der Vernichtung Lebensunwerten Lebens*, Leipzig: Felix Meiner Verlag.

35 Hoche was opposed to the state-controlled 'euthanasia program', particularly as a close relative was one of its victims. See Burleigh (2000), ibid, p.16

36 Binding and Hoche, ibid, p.29

37 Binding and Hoche, op cit, p.31

38 E Häckel (1868), *Natürliche Schöpfungsgeschichte*, Berlin: G. Reimer.

39 S Gould (1992), 'Roots: Ontogeny and phylogeny–revisited and reunited', *Bioessays*, 14:4, pp.275–279

40 S Gould (1992), ibid.

41 These ideas were discussed in Plato, *Republic*, trans. R Waterfield (Oxford: Oxford World's Classics, 1994).

42 F Galton (1883), *Inquiries into Human Faculty and Its Development*, London: Macmillan.

43 For a more detailed account of American eugenics, see E Black (2003), *War against the Weak: Eugenics and America's Campaign to Create a Master Race*, New York: Basic Books; S Selden (2005), 'Transforming Better Babies into Fitter Families: Archival Resources and the History of the American Eugenics Movement 1908–1930', *American Philosophical Society*, 149: pp.199–225

44 R Kluchin (2009), *Fit to Be Tied: Sterilization and Reproductive Rights in America 1950–1980*, New Brunswick: Rutgers University Press.

45 J Lawrence (2000), 'The Indian Health Service and the Sterilization of Native American Women', *The American Indian Quarterly*, 24: pp.400–419

46 In Buck-v-Bell the US Supreme Court ruled 8-1 that an application by Dr John H. Bell, the superintendent at the Virginia State Colony for Epileptics and Feebleminded, to sterilize Carrie Buck, was not in violation of the 14th Amendment of the US Constitution (equal protection). This case paved the way for large scale sterilization. For a comprehensive account of Buck-v-Bell and the US eugenics movement see A Cohen (2006), ibid.

47 C Foster-Kennedy (1942), 'The Problem of Social Control of the Congenital Defective: Education, Sterilization, Euthanasia', *American Journal of Psychiatry*, 99: pp.13–16

48 E Black (2003), ibid.

49 H Friedlander (1995), op cit. p.19

50 E Meltzer (1925), *Das Problem Der Abkürzung 'Lebensunwerten' Lebens*, Halle: Verlag.

51 R Evans (2006), *The Third Reich in Power, 1933-1939*, London: Penguin, p.265

52 S Erapal (2013), 'Diversity Destroyed - Berlin 1933-38,' ed. Deutsches Historiches Museum.

53 RM = Reichsmark. In 1940 1 RM was the equivalent of 2.5 US dollars.

54 Oberöstereiches Landsarchiv (OÖLA) 'Hartheim Statistics' Statl/16a; Statl/18a; Statl/6a. These figures are quoted from a document known as the 'Hartheim Statistics' after a copy of a 39-page leaflet prepared by the Nazi regime to justify its 'euthanasia' program was found at the site of the former Hartheim killing centre by a US Army Officer, Major Charles Dameron in late 1945.

55 Of the US sterilisation laws, the Californian law was the most extensive. The 1917 modification of the original 1909 law enabled the state to enforce sterilisation on anyone deemed undesirable, regardless of diagnosis. This included prisoners, 'sexually promiscuous' people, or any other apparent 'social undesirable'. There was no appeal process and non-Whites were overrepresented in the process on a per capita basis. The last sterilisation in California under this law was in 1963. See A Stern (2005), *Eugenic Nation: Faults and Frontiers of Better Breeding in Modern America*, Berkeley: University of California Press.

56 A Ley (2004), *Zwangssterilisation Und Ärzteschaft. Hintergründe Und Ziele Ärztlichen Handelns 1934-1945*, Frankfurt: Campus Verlag.

57 G Bock (1986), *Zwangssterilisation im Nationalsozialismus. Studien zur Rassenpolitik und Frauenpolitik*, Opladen: Westdeutscher Verlag.

58 Despite the harshness of German sterilisation law, some observers at the time, such as prominent American eugenicist Lothrop Stoddart, viewed its application as moderate by comparison to similar statutes in California. Stoddart visited Germany in 1939 and deemed the German program too conservative. See S Kühl (1994), *The Nazi Connection - Eugenics, American Racism, and German National Socialism*, New York: Oxford.

59 A Ley (2004), ibid.

60 B Müller-Hill (1998), *Murderous Science: Elimination by Scientific Selection of Jews, Gypsies, and Others in Germany, 1933-1945*, G Fraser (trans), Plainview, NY: Cold Spring Harbor Laboratory Press.

61 B Müller-Hill (1998), ibid.

62 The introduction of the Nuremberg laws in 1935 prohibited intermarriage and instances of miscegenation between 'Aryans' and others.

63 For a more detailed account, see R Pommerin (2003), *Sterilisierung der Rheinlandbastarde. Das Schicksal einer farbigen deutschen Minderheit 1918-1937*, Düsseldorf: Droste; B Müller-Hill (1998), op cit; F Carr (2003), *Germany's Black Holocaust, 1890-1945*, Kearney, NE: Morris Publishing.

Endnotes

64 J London (1914), *Told in the Drooling Ward*: https://americanliterature.com/author/jack-london/short-story/told-in-the-drooling-ward

65 F Powell (1912), 'The Care and Training Of Feeble-Minded Children' - *A Presentation at the National Conference of Charities and Correction, at the Fourteenth Annual Session Held In Omaha, Neb., August 25-31*, 1887: http://socialwelfare.library.vcu.edu/issues/care-and-training-of-feeble-minded-children-1887/

66 H Goddard (1912), *The Kallikak family: A study in the heredity of feeble mindedness*, New York: MacMillan.

67 A 'proband' is the first recognised instance of a particular trait under study by geneticists.

68 This refers to the theory of inheritance described by nineteenth century Austrian monk Gregor Mendel whose laws of inheritance formed the basis of early genetics.

69 A Ley and A Hinz-Wessels (2012), ibid, p.47

70 See H Silverman and K Krenzel (1964), 'Alfred Binet: Prolific Pioneer in Psychology', *The Psychiatric Quarterly* (1964) Supplement, 38: pp.323–335

71 See L Zenderland (1998), *Measuring Minds: Henry Herbert Goddard and the origins of American intelligence testing*, New York: Cambridge University Press.

72 A Bashford and P Levine (2010), *The Oxford Handbook of the History of Eugenics*, New York: Oxford University Press.

73 See The Human Rights and Equal Opportunity Commission, Bringing Them Home: The Stolen Generation Report (1997): https://www.humanrights.gov.au/publications/bringing-them-home-stolen-children-report-1997

74 R Jones (2011), 'Eugenics in Australia: The secret of Melbourne's elite', *The Conversation*: https://theconversation.com/eugenics-in-australia-the-secret-of-melbournes-elite-3350 [accessed 22 June 2016]

75 M Snyderman, R Herrnstein (1983), 'Intelligence tests and the Immigration Act of 1924', *American Psychologist*, 38: pp.986–995

76 U Schmidt (2007), *Karl Brandt: The Nazi Doctor*, London: Hambledon Continuum.

77 At that time the Kretschmar baby was referred to as the 'Knauer baby'. For a detailed account see Friedlander (1995), ibid.

78 Richard Kretschmar would survive the Nazi Era and, according to an unsourced literary reference, as late as the 1970s remained steadfast in his opinion about the manner of Gerhard's death, stating that it had allowed he and Lina to 'have other perfectly healthy children of whom the Reich would be proud'. Quoted in D Mainardi (2014), *The Fall: A Father's Memoir in 424 Steps*, M Costa (trans), London: Harvill Seeker, p.36

CHAPTER 3
AKTION T4 – IMPLEMENTATION AND RESISTANCE

79 A Hinz-Wessels (2015), *Tiergartenstraße 4. Schaltzentrale der Nationalsozialistischen 'Euthanasie'-Morde*, Berlin Verlag.

80 H Friedlander (1995), op cit.

81 Despite the infamy of the grey buses, in the early stages of *Aktion* T4 the buses retained their original red colour. *Gekrat* also made use of trains to transport victims from their institutions to *Zwischenanstalten* where they were held until their transport to killing centres. The rail system would later be used with devastating effect.

82 Friedlander (2015), op cit; and Lifton (1986), op cit.

83 Ley (2004), ibid.

84 Patientenarchiv im Christophsbad, Göppingen, Germany.

85 C Koonz (2002), *The Nazi Conscience*, Cambridge Mass: Belknap.

86 '60,000: Nationalsozialistisches Propagandaplakat zur Akzeptanzbereitung für Eugenik und Euthanasie Offsetdruck,' Deutsches Historisches Museum: https://www.dhm.de/lemo/bestand/objekt/pli02843 [Accessed 5 September 2016]

87 B Hinz (1979), *Art in the Third Reich*, New York: Random House.

88 S Kühl (1994), ibid.

89 These films are considered in some detail in a 1991 documentary *Selling Murder: the killing films of the Third Reich* produced by Stewart Lansley, Joanna Mack, and Michael Burleigh for the UK's Channel Four.

Endnotes

90 H Boberach (1968), *Meldungen aus dem Reich 1938 – 1945 Die geheimen Lageberichte des Sicherheitsdienstes der SS*, Munich Verlag, vol. 9: p.3175

91 N Stargardt (2015), *The German War*, London: The Bodley Head.

92 For a detailed consideration of this aspect see Chapter 5 of M Burleigh (1995), *Death and Deliverance: 'Euthanasia' in Germany*, c.1900 to 1945, London: Cambridge University Press.

93 This is a contentious debate. See H Wollasch (1979), *Beiträge Zur Geschichte Der Deutschen Caritas in Der Zeit Der Weltkriege*, Freiburg im Bresigau: Verlag.

94 E Klee (1985), op cit.

95 E Klee (1985), ibid.

96 Encyclical of Pope Pius XII '*Mystici corporis Christi*' 29 June 1943, Paragraph 94: http://w2.vatican.va/content/pius-xii/en/encyclicals/documents/hf_p-xii_enc_29061943_mystici-corporis-christi.html [Accessed 5 September 2016]

97 J Bendersky (2007), *A Concise History of Nazi Germany*, Lanham MA: Rowman and Littlefield, p.147

98 M Burleigh (2002), op cit.

99 Brack was chief of Office 2 in the KdF, which handled matters concerning the Reich Ministries, armed forces, Nazi Party, clemency petitions and complaints received by the Führer from all parts of Germany. He was tasked, along with Karl Brandt and KdF head Philip Bouhler, with the organisation of the 'euthanasia' program.

100 M Burleigh, (2000), *The Third Reich; A New History*, London, Pan Macmillan, p.401. See also, M Burleigh (1994), 'Between Enthusiasm, Compliance and Protest: The Churches, Eugenics and the Nazi 'Euthanasia' Programme', *Contemporary European History*, 3: pp.253–263

101 See M Burleigh (2002), op cit; and I Kershaw (2000), *The Nazi Dictatorship: Problems and Perspectives of Interpretation* 4th Ed, New York: Oxford.

102 N Stargardt (2015), ibid.

103 See E Griech-Polell (2002), *Bishop Von Galen: German Catholicism and National Socialism*, New Haven CT: Yale.

104 A Gill (1994), *An Honourable Defeat; a History of the German Resistance to Hitler*, London: Heinemann.

105 R Krieg (2004), *Catholic Theologians in Nazi Germany*, New York: Continuum.

106 C von Galen (2006), *Bishop Clemens August Graf Von Galen - Files, Letters and Sermons from 1933 to 1946* 2nd Edition, Paderborn: Ferdinand Schöningh.

107 P Löffler (1996), *Bischof Clemens August Graf von Galen – Akten, Briefe und Predigten 1933–1946* 2nd Edition, Paderborn: Ferdinand Schöningh.

108 M Burleigh (2000), op cit.

109 See I Scholl (1983), *The White Rose: Munich, 1942–1943*, A Schultz (trans), Middletown, CT: Wesleyan University Press.

110 See N Cameron and R Stevens (trans) (2000), *Hitler's Table Talk 1941–1944*, London: Enigma Books, p.90

111 A Ley (2009), 'Vom Krankenmord zum Genozid. Die '*Aktion* 14f13 in den Konzentrationslagern', *Dachauer Hefte*, 25.

112 A Ley and A Hinz-Wessels (2012), op cit.

113 M Bazyler (2011), 'The Thousand Year Reich's over one thousand anti-Jewish laws', *The Routledge History of the Holocaust*, J Friedman (ed), Oxford: Routledge. For a German-language biography of Kreyssig, see K Weiß (1988), *Lothar Kreyssig: Prophet der Versöhnung*, Gerlingen: Bleicher Verlag.

114 J Meyer-Lindberg (1991), 'The Holocaust and German psychiatry', *British Journal of Psychiatry* 159: pp.7–12

115 Hessiche Hauptstaatsarchiv, Weisbaden 631a/359 - testimony Dr Gebhard Ritter, May 1948, p.1

116 Bonhoeffer's role in resistance to the T4 program is the subject of debate. See U Gerrens (2001), 'Psychiater unter der NS-Diktatur. Karl Bonhoeffers Einsatz fur rassisch und politisch verfolgte Kolleginnen und Kollegen', *Fortschr Neurol Psychiatr*, 69: p.7

117 M Ströhle et al (2008), Karl Bonhoeffer (1868–1948), *American Journal of Psychiatry*, 165: pp.575–576. See also A Ley (2009) op cit.

118 Lifton (1985), op cit, p.82

119 W Brautigam and C Teller (1993), 'In memoriam John Rittmeister - Nervenarzt, psychoanalytischer Therapeut, Widerstandskampfer', *Nervenarzt*, 64: p.5

120 F Schneider and V Roelcke (2013), 'Psychiatrists in National Socialism', *Der Nervenarzt* 84: pp.1041–1042. According to Dr Johannes Tuchel, the director of the Gedenkstätte Deutscher Widerstand (German Resistance Memorial Centre), it is not clear whether Rittmeister was directly involved in resisting 'euthanasia' as against the anti-Nazi sentiments in his AGIS leaflets.

121 See: John Rittmeister, Gedenkstätte Deutscher Widerstand, Berlin: http://www.gdwberlin.de/en/recess/biographies/index_of_persons/biographie/view-bio/john-rittmeister/?no_cache=1 [Accessed 26 Aug 2016]

122 H Schmuhl (2013), 'Walter Creutz and 'Euthanasia' in the Rhein Province', *Der Nervenarzt* 84(9).

123 C Beyer (2013), 'Gottfried Ewald and the 'Operation T4' in Göttingen', Nervenarzt 84(9).

124 Diary of Dr Ernst Arlt, Steiermärkisches Landesarchiv – Styrian provincial archive 1 1, Tagebuch

125 See M Burleigh (2000), op cit.

126 E Klee (1985), op cit.

127 M Berenbaum (2016), 'T4 Program- Nazi Policy', *Encyclopaedia Britannica Online*. Encyclopaedia Britannica Inc: https://www.britannica.com/event/T4-Program [Accessed 3 October 2016]

128 V Klemperer (1998), *I Shall Bear Witness: The Diaries of Victor Klemperer, 1933-41*, M Chalmers (trans), London: Weidenfeld & Nicolson.

129 Stargardt (2015), op cit.

130 W Shirer (1941), 'Inside Wartime Germany - Part II' Life, February 10, 1941; W Shirer (1941),'Mercy Deaths' in Germany,' *Reader's Digest 6*: pp.55–58; W Shirer (1941), *Berlin Diary: The Journal of a Foreign Correspondent, 1934–1941*, New York: Knopf.

131 See T Noack (2016), 'William L. Shirer and International Awareness of the Nazi 'Euthanasia' Program', *Holocaust and Genocide Studies*, 30: pp.433–457

132 H Gallagher (2001), 'What the Nazi 'Euthanasia Program' Can Tell Us About Disability Oppression', *Journal of Disability Policy Studies*, 12: pp.96-99

133 This story is based upon a second-hand account provided by Frederich Mennecke during his criminal trial, see G Sereny (1983), *Into That Darkness: An Examination of Conscience*, New York: Vintage.

134 Frau Else von Löwis to Frau Buchl 25th November 1940. NMT document N0-001.

135 Heinrich Himmler to Viktor Brack 19 December 1940. NMT document N0-018.

136 H Freidlander (1995), op cit.

137 *Einzelfällen Kost* or 'individual diet'. Faulthauser also murdered patients with injections of morphine and scopolamine (a paralysis agent used in anaesthesia).

138 M Burleigh (2002), op cit.

139 M Burleigh (2002), ibid.

140 B McFarland-Icke (1999), *Nurses in Nazi Germany: Moral Choice in History*, Princeton: Princeton University Press, p.214

141 S Benedict et al (2007) 'Duty and 'Euthanasia': The nurses of Meseritz-Obrawalde', *Nursing Ethics* 14(6): pp.781–794

142 A Ley (2009), op cit.

CHAPTER 4
'DR SCHNEIDER'

143 A Ley and A Hinz-Wessels (2012), op cit.

144 M Burleigh (1997), *Ethics and Extermination – Reflections on Nazi Genocide*, New York: Cambridge, p.124

145 H Friedlander (1995), op cit.

146 A Ley (2009), op cit.

147 H Friedlander (2002), 'Physicians as Killers in Nazi Germany', *Medicine and Medical Ethics in Nazi Germany*, F Nicosia and J Huener (eds), New York: Berghahn Books, pp.59–76

Endnotes

148 M Grabher (2006), *Irmfried Eberl. Euthanasie'-Arzt Und Kommandant Von Treblinka*, Frankfurt am Main: Europäischer Verlag der Wissenschaft.

149 R Lifton (1986), op cit.

150 A Mitscherlich and F Mielke (1960), *Medizin ohne Menschlichkeit: Dokumente des Nürnberger Ärzteprozesses*, Frankfurt am Main: Fischer Bucherei.

151 R Strous (2008), 'Extermination of the Jewish Mentally-Ill During the Nazi Era-the 'Doubly Cursed", Israeli Journal of Psychiatry and Related Sciences, 45: p.4

152 A Ley (2009), op cit.

153 A Ley (2009), ibid.

154 S Schmitt et al (2009), '*Aktion* T4 in the Reich District of Sudetenland on Examples of the District Lunatic Asylums in Sternberg [Šternberk na Moravě], Troppau [Opava] and Wiesengrund [Dobčany] Based on the File R179 of the Bundesarchives Berlin, 1939—1941', *The National Socialist 'Euthanasia' and its victims in the area of the Present Czech Republic 1939-45*, Lern und Gedenkort Schloss Hartheim: http://www.schloss-hartheim.at/projekt-sudetenland-protektorat/en/project-results.htm#3

155 Sonia's family later concealed her fate from their descendants. Her murder remained a family secret until her grandson discovered the circumstances of her death in 2015. In March 2016 about 20 family members travelled to Hamburg in order to lay a *Stolperstein* ('stumbling stone') memorial for Sonia in front of the former family home in remembrance of her.

156 'Neurasthenic' was a term used in the diagnosis of psychosomatic and related complaints.

157 Budapest born playwright and journalist Theodore Herzl became the public face and main advocate for the establishment of a Jewish state through his books *Der Judenstaat* (1896) and *Alteneueland* (1902). Herzl died age 44, when Margarethe was a child. Among the excellent biographies of Herzl are E Pawel (1989), *The Labyrinth of Exile - A Life of Theodor Herzl*, New York: Farrar, Straus & Giroux; and A Elon (1975), *Herzl*, New York: Holt, Rinehart and Winston.

158 E Pawel (1989), ibid. p.563

159 Clinical file of Margarethe Neumann - *Dokumentationsarchiv des österreichischen Widerstandes*.

160 For a comprehensive account of the ghetto at Theresienstadt, see H Adler (2017), *Theresienstadt 1941-1945: The Face of a Coerced Community*, New York: Cambridge University Press.

161 V Frankl (2006), *Man's Search for Meaning*, Boston: Beacon Press.

162 K Kwiet (1984), 'The Ultimate Refuge: Suicide in the Jewish Community under the Nazis', *Leo Baeck Institute Yearbook*. 29: pp.135–167

163 See A Hájková (2018), 'Medicine in Theresienstadt', *Social History of Medicine*: https://doi-org.ezproxy1.library.usyd.edu.au/10.1093/shm/hky066

164 P Klein (2005), 'Theresienstadt: Ghetto oder Konzentrationslager?', *Theresienstädter Studien und Dokumente*, pp.111–123

165 See R Mueller (1991), *Hitler's Ostkrieg und die deutsche Siedlungspolitik: Die Zusammenarbeit von Wehrmacht, Wirtschaft und SS*, Frankfurt/Main: Fischer.

166 A Weale (2010), *The SS - A New History*, London: Little Brown.

167 Affidavit of Otto Ohlendorf, 5 November 1945 (PS 2620), in IMT Trial, Prosecution Documents, roll 11, frames 0044-0045.

168 Deposition by Karl Wolff, *Staats Archiv München*, Az.10a Js 39/60.

169 C Browning (2004), *The Origins of the Final Solution - the Evolution of Nazi Jewish Policy*, September 1939-March 1942, Lincoln: University of Nebraska Press.

170 C Browning (2004), ibid, p.283

171 R Hilberg (1961), *The Destruction of the European Jews*, London: W H Allen, p.218

172 R Lifton (1985), ibid, p.15

173 Goldensohn had served as a psychiatrist with the US Army's 63rd Division and was tasked with monitoring the mental health of the 21 defendants in the main Nuremberg War Crimes trial.

174 L Goldensohn (2004), *The Nuremburg Interviews*, New York: Vintage, p.390. Ohlendorf was sentenced to death in August 1948 after his conviction in the so-called 'Einsatzgruppen Trial' (United States of America vs Otto Ohlendorf et al).

175 R MacNair (2005), *Perpetration-Induced Traumatic Stress: The Psychological Consequences of Killing*, New York: Authors Choice Press, p.46

176 H Höhne (1969), *The Order of the Death's Head: The Story of Hitler's SS*, London: Penguin, p.363

177 Nuremburg Military Tribunal Document NO-205 'Letter to Heinrich Himmler concerning the x-ray sterilization proposal', 23 June 1942: http://nuremberg.law.harvard.edu/documents/1252-letter-to-heinrich-himmler?q=NO-205#p.1

178 P Longerich (2010), *Holocaust:The Nazi Persecution and Murder of the Jews*, Oxford: Oxford University Press.

179 Heydrich is one of the most sinister characters of the Nazi regime. His actions in the Holocaust and as Protector of Bohemia and Moravia, his assassination in 1942 and the retribution for his death wrought on the village of Lidice, are well beyond the scope of this book. There are many excellent biographies of Heydrich including M Dederichs (2009), *Heydrich: The Face of Evil*, Drexel Hill PA: Casemate; and R Gerwarth (2011), *Hitler's Hangman: The Life of Heydrich*, London: Yale University Press.

180 Göring to Heydrich, July 31, 1941: listed in L Dawidowicz (1976), *A Holocaust Reader*, West Orange: Behrman, pp.72–73

181 A Ley (2009), op cit, pp.36–49

182 See S Friedländer (2007), *The Years of Extermination: Nazi Germany and the Jews 1939-1945*, New York: Harper Collins.

183 Y Arad (1987), *Belzec, Sobibor, Treblinka: The Operation Reinhard Death Camps*, Bloomington, Il: Indiana University Press.

184 The notoriously sadistic guard at Treblinka, Ivan Marchenko, also known as 'Ivan the Terrible', was a Hiwi, as was John Demjanjuk, the former car worker from Cleveland Ohio, extradited and convicted in 2011 by a German court as an accessory to the murder of 28,060 Jews while acting as a guard at the Sobibór concentration/extermination camp.

185 P Longerich (2010), ibid.

186 Y Arad (1987), ibid.

187 The inclusion of the concentration/extermination camp in Lublin (Majdanek) as a Reinhard camp is debated. In 2000, researchers chanced upon a document in the Public Record Office in Kew that came to be known as 'Höfle Telegram' (HW 16/23, decode GPDD 355). This was a decoded intercepted transmission to Eichmann from *SS Sturmbannführer* Hermann Höfle on 11 January 1943.

The transmission listed data from *Deutsche Reichsbahn* (German Railways) that indicated that more than 1.2 million people had been killed in four camps, including Majdanek, by 31 December 1942, creating the impression that Majdanek may have been a Reinhard camp. See P Witte and S Tyas (2001), 'A New Document on the Deportation and Murder of Jews during Einsatz Reinhard 1942', *Holocaust and Genocide Studies* 15: p.472

188   G Aly (1999), *'Final Solution': Nazi Population Policy and the Murder of the European Jews*, B Cooper and A Brown (trans), London: Arnold.

189   According to Holocaust historian Konrad Kwiet, it is debatable whether this massacre was perpetrated by Lange's men or other SS formations such as members of a *SS Totenkopfstandarte* (Death's Head Regiment) and of *SS-Einsatzkommando*.

190   P Longerich (2010), op cit.

191   Deposition by H Renfranz, dated 10.10.1962, StA Hannover, Az. 2 Js 614/62 [ZSL, Az. V 203 AR-1101/1960, Bl2] cited in E Klee (1983), *'Euthanasie' im NS-Staat. Die 'Vernichtung lebenswerten Lebens'*, Frankfurt a.M. Verlag 1983, pp.106 and 190

192   C Browning (2004), *The Origins of the Final Solution*, London: Arrow Books.

193   Letter from Becker to Rauff. 16th May 1942, IMT-Doc. PO-501-PS.

194   Organization Todt was headed by the engineer Fritz Todt. Before the war, Todt's group mainly constructed highways (*Autobahnen*).

195   See P Heberer (2011), 'Von der *Aktion* T4 zum Massenmord an den europäischen Juden', in A Ley, G Morsch and B Perz (eds), *Neue Studien zu Nationalsozialistischen Massentötungen durch Giftgas*, Berlin: Metropol: pp.165–175

196   M Bryant (2014), *Eyewitness to Genocide – The Operation Reinhart Death Camp Trials 1955-66*, Knoxville: University of Tennessee Press.

197   M Bryant (2014), ibid.

198   Franz was to become the last commandant at Treblinka. His sadism was notorious, including his encouragement of his large pet dog 'Bari' to attack prisoners.

199   M Grabher (2006), ibid.

200   Irmfried Eberl to Ruth Eberl 30 July 1942 Hessiches Hauptstaatsarchiv, Wiesbaden, 631a, 1631 (III/683/6, 149-150).

201   R Lifton (1986), op cit.

202 Wirth insisted that the Reinhard camps use his method of engine exhaust, despite the frequent problems with engine breakdown, see Y Arad (1987), op cit.

203 I Baxter (2016), *The SS of Treblinka*, Stroud: The History Press.

204 Mätzig had worked, presumably with Eberl, at Brandenburg and was one of the many T4 personnel transferred to Reinhard camps. See E Klee, W Dressen and V Riess (1988), *The Good Old Days: The Holocaust as Seen by Its Perpetrators and Bystanders*, D Burnstone (trans), Old Saybrook, CT: Konecky & Konecky, p.245

205 Wirth was killed by partisans in May 1944 in Fiume (present day Rijeka in Croatia).

206 G Sereny (1983), op cit, p54. Stangl was captured by the US Army after the war and later escaped to Syria with the assistance of the 'Ratline', a network of Catholic clergy sympathetic to the former Nazi regime. He later fled to Brazil. He was apprehended by Brazilian police in 1967 after being identified by famous Nazi hunter Simon Wiesenthal. A court in Düsseldorf sentenced him to life imprisonment for his role in the deaths of 900,000 people during *Aktion* T4 and *Aktion* Reinhard. Throughout 1971, Gitta Sereny conducted nearly 70 hours of interview with the imprisoned Stangl in preparation for her book *Into That Darkness*.

207 G Sereny (2002), *The Healing Wound -- Reflections on Germany 1938-2001*, New York: Norton, p117

208 Y Arad (1987), op cit.

209 Fritz Bleich, a witness at the Nuremberg International Military Tribunal, places Eberl at Auschwitz I for a period of 6 months in 1943. IMT Nuremberg NO-860 'Statement of Fritz Bleich'.

210 Many *Aktion* T4 and *Aktion* Reinhard staff were deployed to anti-partisan operations near Trieste on the border between Italy and Slovenia. A small-scale *Polizeihaftlager* (Police detention camp), was established in a disused rice husking plant at San Sabba and Erwin Lambert constructed a large crematorium in the centre. Around 3000 people were murdered at *Riseria San Sabba*.

211 E Klee (2004), *Was sie taten – was sie wurden. Ärzte, Juristen und andere Beteiligte am Kranken-oder Judenmord*, Frankfurt am Main: Fischer-Taschenbuch-Verlag.

212 M Grabher (2006), op cit.

213 E Kogon (1946), Der SS-Staat. *Das System der Deutschen Konzentrationslager*, Munich: Karl Alber.

214 H Friedlander (2002), ibid; M Dudley and F Gale (2002), 'Psychiatrists as a Moral Community? Psychiatry under the Nazis and Its Contemporary Relevance', *Australian and New Zealand Journal of Psychiatry* 36: pp.585–594

215 M Bryant (2005), *Confronting the 'Good Death': Nazi Euthanasia on Trial, 1945–1953*, Boulder: University Press of Colorado.

216 D Deutsch (2016), "Immer Mit Liebe': Empathic Violence in Nazi Euthanasia', *Holocaust Studies*, 22: p.1

217 S Benedict and J Kuhla (1999), 'Nurses' Participation in the Euthanasia Programs of Nazi Germany', *Western Journal of Nursing Research* 21: p.2

218 G Hohendorf (2013), *Der Todals Erlösungvom Leiden*, Göttingen: Wallstein Verlag.

CHAPTER 5
'THERE WAS WONDERFUL MATERIAL AMONG THOSE BRAINS'

219 A Ley and A Hinz-Wessels (2012), op cit, p.109

220 Brandenburg T4 doctor Irmfried Eberl's diary includes a note to collect him by car on that date. See A Ley and A Hinz-Wessels (2012), op cit, p.106

221 J Hallervorden (1948), 'Oligodendrogliom Nach Hirntrauma', *Nervenarzt* 19.

222 S Coskun, et al (2011), 'Post-Traumatic Glioblastoma Multiforme: A Case Report', *Eurasian J Med*, 43: pp.50–53

223 W Krücke (1966), 'Julius Hallervorden - 1882-1965', *Acta Neuropathologica*, 6: pp.113–116

224 J Hallervorden and H Spatz (1922), 'Eigenartige Erkrankung im extrapyramidalen System mit besonderer Beteiligung des Globus pallidus und der Substantia nigra', *Zeitschrift für die gesamte Neurologie und Psychiatrie*, 79: pp.254–302

225 E Richardson (1990), 'Julius Hallervorden', in *Founders of child neurology*, S Ashwal (ed), San Francisco: Norman Publishing, pp.506–512

226 In 2001 the condition was renamed Pantothenate kinase-associated neurodegeneration (PKAN) - B Zhou, et al (2001), 'A novel pantothenate kinase

gene (PANK2) is defective in Hallervorden-Spatz syndrome', *Nature, Genetics*, 28: pp.345–349

227 W. Seidelman (2002), 'Pathology of Memory:German Medical Science and the Crimes of the Third Reich', *Medicine and Medical Ethics in Nazu Germany*, F Nicosia and J Huener (ed), New York: Berghahn Books.

228 For a detailed account, see: H. Schmuhl (2009), 'Brain Research and the Murder of the Sick: The Kaiser Wilhelm Institute for Brain Research 1937-1945', *The Kaiser Wilhelm Society under National Socialism*, S Heim, C Sachse and M Walker (eds), New York: Cambridge University Press, pp.99–119

229 W Haymaker (1951), 'Cecile and Oskar Vogt, on the Occasion of Her 75th and His 80th Birthday', *Neurology* :1, pp.179–204

230 H Schmuhl (2009), ibid.

231 A Ley and A Hinz Wessels (2012), op cit, p.174, 176.

232 Ref: IVb 3088/39 - 1079 Mi. In E Klee (1985), op cit, p.80

233 Circular by the Minister of the Interior on 20 September 1941 Az.: IVb-1981/41-1079 Mi, in E Klee (1985) op cit, p.303

234 H Heinze: *Vorschläge für eine zukünftige Raumgestaltung jugend-psychiatrischer Anstalten*. 6 February 1942, in E Klee (1985), op cit, pp.380-381

235 T Beddies (2004), 'Kinder-'Euthanasie' in Berlin-Brandenburg', *Dokumente Zur Psychiatrie Im Nationalsozialismus*, T Beddies and K Hübener (eds), Berlin-Brandenburg: Be.Bra Wissenschaft Verlag.

236 E Sheffer (2017), *Asperger's Children*, New York: Norton.

237 L Pelz (2003), 'Kinderärzte im Netz der 'NS-Kindereuthanasie' am Beispiel der 'Kinderfachabteilung' Görden', *Monatsschrift Kinderheilkunde*,10: pp.1027–1032

238 T Beddies (2004), ibid.

239 D Roer (1992), 'Psychiatrie in Deutschland, 1933-1945: Ihr Beitrag zur 'Endlösung der sozialen Frage' am Beispiel der Heilanstalt Uchtspringe', *Psychologie und Gesellschaftskritik*, 16: pp.15–37

240 K Alt (2003), 'Über ländliche Beschäftigung der Kransinnigen in Anstalts- und Familienpflege, in Uchtspringer Schriften zur Psychiatrie, Neurologie, Schlafmedizin', *Psychologie und Psychoanalyse*, 1: pp.153–165

241 U Hoffmann (2005), 'Verlegt auf Anweisung des Reichsverteidigungskommissars' - Zu den Transporten aus der Landesheilanstalt Uchtspringe in die 'Euthanasie' - Anstalten Brandenburg und Bernburg 1940-1941', *Uchtspringer Schriften zur Psychiatrie, Neurologie, Schlafmedizin, Psychologie und Psychoanalyse*, 3: pp.69–83

242 F Thomas, et al (2006), 'A Cold Wind Coming: Heinrich Gross and Child Euthanasia in Vienna', *Journal of Child Neurology*, 21.

243 V Frankl (2006), ibid, pp.154–155

244 F Thomas, et al (2006), ibid.

245 H Czech (2002), 'Forschen ohne Skrupel. Die wissenschaftliche Verwertung von Opfern der NS-Psychiatriemorde in Wien', in E Gabriel, W Neubauer (eds), *Zur Geschichte der NS-Euthanasie in Wien: Von der Zwangssterilisation zur Ermordung*, Vienna: Böhlau Verlag, pp.147–187

246 Bunke only gained his medical degree in April 1941.

247 J Peiffer (1999), 'Assessing Neuropathological Research Carried out on Victims of the 'Euthanasia' programme: With Two Lists of Publications from Institutes in Berlin, Munich and Hamburg', *Medizinhistorisches Journal*, 34: pp.339–355

248 Ley (2009), ibid, p.106

249 Report Prof. Carl Schneider from January, 24th 1944 'Über Stand, Möglichkeiten und Ziele der Forschung an Idioten und Epileptikern im Rahmen der *Aktion*', *Bundesarchiv Koblenz* (Fc 1807 P, Roll 12).

250 H. Friedlander (1995), op cit. p.343

251 H Schmuhl (2009), op cit.

252 J Peiffer (1999), ibid.

253 Bundesarchiv, Koblenz, R22-4209 cited by M. I. Shevell and J. Peiffer (2001), 'Julius Hallervorden's Wartime Activities: Implications for Science under Dictatorship,' *Pediatr Neurol* 25, no. 2.

254 J Peiffer (1991), 'Neuropathology in the Third Reich. Memorial to Those Victims of National-Socialist Atrocities in Germany Who Were Used by Medical Science', *Brain Pathology*, 1: pp.125–131

255 For a comprehensive account of the Alexander's life and career, see U Schmidt (2004), *Justice at Nuremberg*, London: Palgrave Macmillan.

256 International Military Tribunal (1946), '*Major War Criminals (USA, France, UK, and USSR) v. Hermann Goering et al. 1945-46*': https://www.loc.gov/rr/frd/Military_Law/pdf/NT_Vol-I.pdf [Accessed 7 July 2015]

257 International Military Tribunal (1946), ibid.

258 L Alexander (1946), 'Neuropathology and Neurophysiology Including Electroencephalography in Wartime Germany', *Combined Intelligence Objectives Subcommittee (CIOS) Report*, Item No. 24, File No. Xxvii-I (Document L170).

259 L Alexander (1946), ibid, p.17

260 Hallervorden claimed this was a mistake and wrote to Alexander seeking to correct this. There is no evidence of Alexander responding - *Archiv zur Geschichte der Max-Planck-Gesellschaft, Berlin*, Hauptabteilung II, Rep. 1 A, Personalia Hallervorden Heft 5, Bl. 4.

261 L Alexander (1949), 'Medical Science Under Dictatorship', *New England Journal of Medicine*, 241: pp.39–47

262 The true nature of Schaltenbrand's activities was not revealed until the 1990s but anecdotal accounts of these experiments were likely known at the time. See M Shevell and B Evans (1994), 'The 'Schaltenbrand Experiment', Wurzburg, 1940: Scientific, Historical, and Ethical Perspectives', *Neurology*, 44:2.

263 J Aarli (2014), *The History of the World Federation of Neurology: The First 50 Years*, London: Oxford University Press.

264 J Peiffer (1997), *Hirnforschung Im Zwielicht. Beispiele Verführbarer Wissenschaft Aus Der Zeit Des Nationalsozialismus*, Abhandlungen Zur Geschichte Der Medizin Und Der Naturwissenschaften, Husum: Matthiesen.

265 U Schmidt (2004), ibid.

266 L Kaebler (2013), Jewish Children with Disabilities and Nazi 'Euthanasia' Crimes, *Bulletin of the Carolyn and Leonard Miller Centre for Holocaust Studies*, 17.

266 H Czech (2014), 'Abusive medical practices on 'euthanasia' victims in Austria during and after World War II', in S Rubenfeld, S Benedict S (eds), *Human subjects research after the Holocaust, Cham*: Springe International Publishing, pp.109–126. Fredericke Pusch was never prosecuted and enjoyed a productive post-war career.

267 A Mitscherlich and F Mielke (2009), *Medizin Ohne Menschlichkeit: Dokumente Des Nürnberger Ärzteprozesses*, Frankfurt: Fischer Taschenbuch Verlag.

268 The Nuremberg Code, 1949. 'Trials of War Criminals before the Nuremberg Military Tribunals under Control Council Law No. 10', Vol. 2, pp.181–182. Washington, D.C.: U.S. Government Printing Office.

269 R Carlson, et al (2004), 'The revision of the Declaration of Helsinki: past, present and future', *British Journal of Clinical Pharmacology*, 57: pp.695–713. Shuster writes: 'The Nuremberg Code has not been officially adopted in its entirety as law by any nation or as ethics by any major medical association. Nonetheless, its influence on global human-rights law and medical ethics has been profound'. See E Shuster (1997), 'Fifty Years Later: The Significance of the Nuremberg Code', *New England Journal of Medicine*, 337: pp.1436–1440

270 H Beecher (1996), 'Ethics and Clinical Research', *New England Journal of Medicine*, 174: pp.1354–1360

271 D. Kondziella (2009), 'Thirty Neurological Eponyms Associated with the Nazi Era', *European Neurology*, 62:1.

272 R Strous and M Edelman (2007), 'Eponyms and the Nazi Era: Time to Remember and Time For Change', *Israel Medical Association Journal*, 9: pp.207–214

273 H Czech (2005), 'Dr. Hans Asperger und die 'Kindereuthanasie' in Wien – mögliche Verbindungen', in *Auf den Spuren Hans Aspergers: Fokus Asperger-Syndrom: Gestern, Heute, Morgen*, A Pollak (ed), Stuttgart: Schattauer, pp.24-29

274 H Czech (2018), 'Hans Asperger, National Socialism, and 'race hygiene' in Nazi-era Vienna', *Molecular Autism*, 9-29. DOI: 10.1186/s13229-018-0208-6. eCollection 2018.

275 American Psychiatric Association (2013), *Diagnostic and Statistic Manual of Mental Disorders – Fifth Edition*. Washington APA Press.

276 A Caplan (1992), 'The Doctors' Trial and Analogies to the Holocaust in Contemporary Bioethical Debates', in: *The Nazi Doctors and the Nuremberg Code*, G Annas and M Grodin (eds), New York: Oxford, pp 258-275

277 World Psychiatric Association *Madrid Declaration on Ethical Standards for Psychiatric Practice* made in Madrid on August 25, 1996 http://www.wpanet.org/detail.php?section_id=5&content_id=48 [Accessed 13 June 2015]

278 *Dark Years: The Legacy of Euthanasia* (UK, 2016). Directed and written by Jasmine Wingfield, Stormtree Productions.

279   G Aly (1984), *Bericht*. MPG II. Abt., Rep. 1F, Az A-II-7a Besondere Aufgaben Hirnschnittsammlung.

280   J Wingfield (2016), ibid.

281   *Tübingen University Abschlussbericht der Commission zur Überprüfung der Präparätesammlungen in den medizinischen Einrichtungen der Universität Tübingen im Hinblick auf Opfer des Nationalsozialismus*. MPG II. Abt., Rep. 1F, Az A-II-7a Besondere Aufgaben Hirnschnittsammlung. Jul 13, 1989.

282   P Weindling (2017), 'Exhuming the past: 'euthanasia' victims and other specimens buried by the Max Planck Society and issues arising and additional subjects that have arisen between 1990 and 2017', *The Galilee Second International Workshop on Medicine after the Holocaust* (9 May 2017).

283   For a detailed account of this issue see P Weindling (2012), 'Cleansing Anatomical Collections: The Politics of Removing Specimens from German Anatomical and Medical Collections 1988–92', *Annals of anatomy – Anatomischer Anzeiger : official organ of the Anatomische Gesellschaft*, 194:3.

284   'History of the Kaiser Wilhelm Society under National Socialism' https://www.mpg.de/9811513/kws-under-national-socialism [Accessed 21 September 2016]

285   C Sachse (ed) (2003), *Die Verbindung nach Auschwitz. Biowissenschaftenund Menschenversuche an Kaiser-Wilhelm-Instituten*. Wallstein: Göttingen.

286   C Sachse (2011), 'Apology, responsibility, memory. Coming to terms with Nazi medical crimes: the example of the Max Planck Society', *Eur Arch Psychiatry Clin Neurosci*, 261 (Suppl 2): S202–S206.

287   'The Max Planck Institute for Brain Research commemorates its tragic past' October 29, 2015 https://www.mpg.de/9718719/ [Accessed 12 August 2016]

288   There is now a new project on Brain Research on Nazi victims at the Max Planck Society; https://www.meduniwien.ac.at/web/ueber-uns/news/detailseite/2017/news-im-august-2017/max-planck-gesellschaft-stellt-sich-der-ns-vergangenheit/; http://www.get.med.tum.de/hirnforschung-instituten-der-kaiser-wilhelm-gesellschaft-im-kontext-nationalsozialistischer; http://www.sciencemag.org/news/2017/01/germany-probe-nazi-era-medical-science.

CHAPTER 6
'CURE OR DESTROY'

289 For the authors' prior work on the twentieth century history of psychiatry, and the contemporary significance of the National Socialist period and the Holocaust for the profession, see: M Robertson, E Light, W Lipworth and G Walter (2016), 'The Contemporary Significance of the Holocaust for Australian Psychiatry', *Health and History* 18(2), pp.99-120; and M Robertson, E Light, G Walter, and W Lipworth (2017), 'Psychiatry, Genocide and the National Socialist State: Lessons Learnt, Ignored and Forgotten' in N Marczak & K Shields (eds), *Genocide Perspectives V: A Global Crime, Australian Voices*, (pp. 69-89). Sydney: UTS ePRESS.

290 For a more in-depth analysis of this, see both G Cocks (1997), *Psychotherapy in the Third Reich: The Göring Institute*, New York: Oxford; and J Goggin and E Brockman-Goggin (2001), *Death of a Jewish Science: Psychoanalysis in the Third Reich*, West Lafayette: Purdue University Press.

291 E Engstrom, et al (2006), 'Emil Wilhelm Magnus Georg Kraepelin (1856-1926)', *The American Journal of Psychiatry*, 163:1710

292 J Joseph and N Wetzel (2013), 'Ernst Rudin: Hitler's Racial Hygiene Mastermind', *Journal of the History of Biology*, 46: pp.1–30

293 D Blasius (1994), Einfache Seelenstörung. *Geschichte Der Deutschen Psychiatrie 1800-1945*, Frankfurt am Main: Fischer Verlag.

294 H Faulstich (1998), *Hungersterben in Der Psychiatrie 1914–1949. Mit Einer Topographie Der Ns-Psychiatrie*, Freiburg im Breisgau: Lambertus-Verlag.

295 E Rittershaus (1927), *Die Irrengesetzgebung in Deutschland*, Berlin: De Gruyter.

296 H Schmeidebach and S Priebe (2004), 'Social Psychiatry in Germany in the Twentieth Century: Ideas and Models', *Medical History*, 48(4): pp.449–472

297 A Pernice (1995), 'Family Care and Asylum Psychiatry in the Nineteenth Century: The Controversy in the Allgemeine Zeitschrift Fur Psychiatrie between 1844 and 1902', *History of Psychiatry*, 21: pp.55–68

298 Burleigh argues that enthusiasm for eugenic ideas represented a form of moral panic in the face of such changes. See M Burleigh (2002), op cit, pp.112–127

299 M Burleigh (2002), op cit.

300 E Shorter (1997), *A History of Psychiatry: From the Era of the Asylum to the Age of Prozac*, New York: Wiley.

301 C Tsay (2013), 'Julius Wagner-Jauregg and the Legacy of Malarial Therapy for the Treatment of General Paresis of the Insane', *The Yale Journal of Biology and Medicine*, 86: pp.245–254

302 M Sakel (1938), 'The Pharmacological Shock Treatment of Schizophrenia', *American Journal of Psychiatry*, Suppl. 94.

303 For a detailed account of ECT in Germany in this period see L Rzesnitzek and S Lang (2017), 'Electroshock Therapy in the Third Reich', *Medical History* 61: pp.66–88

304 A Braunmühl (1942), 'Über mobile Elektrodentechnik bei der Elektrokrampftherapie', *Archiv fur Psychiatrie*, 114: pp.605–610

305 W Holzer (1941), 'Über die Methodik der Elektroschocktherapie', *Allgemeine Zeitschrift fur Psychiatrie*, 118: pp.357–379; and W Holzer (1952), 'Uber eine Entwicklungsreihe von Elektoschockgeraten', *Allgemeine Zeitschrift fur Psychiatrie*, 121: pp.124–140

306 A Ley (2006), 'Psychiatriekritik durch Psychiater. Sozialreformerische und professionspolitische Ziele des Erlanger Anstaltsdirektors Gustav Kolb', in Heiner Fangerau and Karen Nolte (eds), *Moderne Anstaltspsychiatrie im 19.und 20. Jahrhundert. Legitimation und Kritik*, Stuttgart: Steiner, pp.195–220

307 H Schmiedebach and S Priebe (2004), ibid.

308 T Stöckle (1998), *Die 'Aktion T4' : die 'Vernichtung Lebensunwerten Lebens' in den Jahren 1940/41 und die Heilanstalt Christophsbad in Göppingen*, http://www.edjewnet.de/aktion_t4/

309 M Burleigh (1997), op cit.

310 A Borthwick et al (2001), 'The Relevance of Moral Treatment to Contemporary Mental Health Care', *Journal of Mental Health*, 10: pp.427–439

311 R Whitaker (2002), *Mad in America: Bad Science, Bad Medicine, and the Enduring Mistreatment of the Mentally Ill*, New York: Perseus.

312 M Burleigh (1997), op cit, p.116

313 This process was outlined in detail by Dr Gerhardt Schmidt who was appointed as the director of the Eglfing-Haar institution near Munich, a notorious site of

'decentralised euthanasie'. See G Schmidt (2012), *Selektion in Der Heilanstalt, 1939-1945*, 2nd ed. Berlin: Springer Verlag.

314 P Nitsche (1902), *Über Gedächtnisstörung in zwei Fällen von organischer Gehirnkrankheit*, Med Diss, Berlin.

315 H. Friedlander, (1995) op cit, p.79

316 B Böhm (2012), 'Paul Nitsche – Reformpsychiater Und Hauptakteur Der NS-'Euthanasie'', *Der Nervenarzt*, 83: pp.293-302

317 D Schreber (1903/2000), *Memories of My Nervous Illness*, I Macalpine and R Hunter (trans), New York: New York Review of Books.

318 Schreber (1903/2000), ibid. pp.139-40

319 Schreber's memoirs have been analysed by many scholars for their content and as an account of psychosis. His story formed the basis of one of Freud's main case studies and study of the psychological defences against troublesome sexual impulses. See S Freud (1911), 'Psycho-analytic notes on an autobiographical account of a case of paranoia (dementia paranoids)', *Standard Edition of the Complete Psychological Works of Sigmund Freud*, Vol. XII, J. Strachey (trans), London: Vintage, pp.3-82

320 These figures are reported in the official history compiled by Boris Böhm - *Die Sonne der Deutschen Psychiatrie ging auf dem Sonnenstein bei Pirna in Sachsen auf. Die Geschichte der Heil- und Pflegeanstalt Sonnenstein 1811–1939*. Kuratorium Gedenkstätte Pirna.

321 B Böhm (2012), ibid.

322 Nitsche was invited that year to write a chapter on his work in asylum reform for a definitive text of psychiatry - 'Allgemeine Therapie und Prophylaxe der Geisteskrankheiten'. In: O Bumke (1929), *Handbuch der Geisteskrankheiten, Bd. IV*, Springer, Berlin, pp.1–131

323 C Koonz (2002), ibid.

324 R Lifton (1986), op cit.

325 R Lifton (1986), op cit. p.34

326 C Koontz (2002), ibid.

327 M Rüther (1997), 'Ärztliches Standeswesen im Nationalsozialismus 1933-1945', in: Robert Jütte (ed.), *Geschichte der deutschen Ärzteschaft. Organisierte Berufs- und*

Gesundheitspolitik im 19. und 20. Jahrhundert, Köln: Deutscher Ärzteverlag, p. 143-193. Many influential positions in the Reich Physicians Chamber were given to ardent Nazis and previous members of paramilitary groups such as the *Freikorps*, the *Sturm Abteilung* (SA) or the SS.

328 M Kater (1989), *Doctors under Hitler*, London: Chapel Hill.

329 O Haque et al (2012), 'Why Did So Many German Doctors Join the Nazi Party Early?, *International Journal of Law and Psychiatry* 35.

330 German psychiatrists were well integrated into the Wehrmacht. Each army district had an advisory psychiatrist (*Beratender Psychiater*) who provided data to the *Oberkommando der Wehrmacht* (OKW-central command) including the incidence of different forms of psychological disturbance, self-inflicted wounds, desertion and insubordination.

331 There is some debate over the significance of bed shortages to the ongoing killing of patients after the closure of the *Aktion* T4 killing centres in late 1941. Historian Götz Aly has argued this process was centralised under the orders of Karl Brandt and overseen by Nitsche. See G Aly (1989), 'Die '*Aktion Brandt*'. Bombenkrieg, Bettenbedarf Und „Euthanasie', in: *Aktion T4, 1939-45:Die „Euthanasie' - Zentrale in Der Tiergartenstrasse 4*, G Aly(ed), Berlin: Edition Hentrich.

332 Friedlander (1995), op cit.

333 Schuman later moved to the Auschwitz complex where he performed sterilization experiments on prisoners. His fellow 'euthanasia physicians' at Pirna were Kurt Borm, Klaus Endruweit, Curt Schmalenbach and Ewald Wortmann (code name 'Dr. Friede').

334 E Klee (1985), op cit.

335 F Schwanninger and P Eigelsberger (2015), 'The National Socialist 'Euthanasia' and Its Victims in the Area of the Present Czech Republic 1939-45,' Dokumentationsstelle Hartheim des OÖ. Landesarchiv, http://www.schloss-hartheim.at/projekt-sudetenland-protektorat/en/project-results.htm.

336 V Klemperer (1998), op cit.

337 Handschriftliche Aussage Nitsche vom 25.03.1946, SHStA Dresden, Staatsanwaltschaft beim Landgericht Dresden 11120/ 2526 Bl. 43a.

338 B Böhm (2012), op cit.

339 A Ley, 'Der Beginn des NS-Krankenmorde in Brandenburg/Havel. Zur Bedeutung der 'Brandenburger Probetötung'' für die *Aktion* T4, in: *Zeitschrift für Geschichtswissenschaft* 58 (2010), pp.321-331

340 See L Rzesnitzek (2017), ibid.

341 His appointment was to a 'front' organization for *Aktion* T4 the *Reichsarbeitsgemeinschaft Heil- und Pflegeanstalten* (Reich Cooperative for State Hospitals and Nursing Homes).

342 Friedlander (1995), op cit. p.70.

343 B Böhm, op cit.

344 H Schmuhl (2009), op cit.

345 This narrative is derived from the Christophsbad file for Georg Mall and the investigative work of Hans-Joachim Lang, 'Töten Sie Ihre Eigenen Bruder' (Killing Your Own Brother) *Die Tageszeitung*, 6/7 December 1997.

346 Clinical file of Georg Mall, *Patient Archiv Christophsbad Heilanstalt.*

347 Pentylenetetrazol ('Cardiazol' or 'Metrazol') is a stimulant medication that when administered at high doses induces seizures. Hungarian psychiatrist Ladislas Meduna noted clinical improvement in patients with both epilepsy and schizophrenia after they had suffered seizures. In 1935 Meduna published a case series of 110 patients showing more than half had recovered after seizures induced by Metrazol. This treatment was preferred to the hazardous insulin coma therapy and malaria therapy and was used extensively throughout the late 1930s and early 1940s. In 1938 Italian psychiatrist Ugo Cerletti developed a method of inducing seizures with electrical stimulation, a treatment now known as 'electroconvulsive therapy' (ECT). For Meduna's original paper see L Meduna (1937), *Die Konvulsionstherapie der Schizophrenie*, Halle, Germany, Carl Marhold. Cerletti's work is reported in U Cerletti (1940), *L'Elettroshock. Rivista Sperimentale di Frenatria*. I, pp.209-310

348 The term 'hebephrenic schizophrenia' ('Hebe' is the Ancient Greek goddess of youth) refers to the form of the illness characterised by the insidious onset of disorganised and childish behaviour, as against 'paranoid schizophrenia' where the clinical picture is dominated by psychotic symptoms such as delusions and hallucinations. 'Hebephrenia' tends to have an earlier onset and a poorer prognosis and is more akin to the description of 'dementia praecox' by Swiss psychiatrist Eugen Bleuler.

349 M Kretschmer (1997), 'The sanatorium Weißenau 1933 to 1945', in E Peter (ed), *Ravensburg in the Third Reich: contributions to the history of the city*, Ravensburg: Oberschwäbische Verl-Anstalt Ravensburg, p.36

350 *National Archives and Records Administration* (NARA) RG338, Microfilm Publication T-1021, Roll 18.

351 H-J Lang (1997), ibid.

352 G Mall (1967), *Das Gesicht des seelisch Kranken*, Konstanz: Schnetztor-Verl.

353 See Chapter 2.

354 Other 'euthanasia trials' took place at Hadamar (1945) and Klagenfurt (1946). For a detailed review of the legal complexities of these trials, see Bryant (2005), ibid.

355 Anklage gegen Nitsche und Anderren -SHStA Dresden, Staatsanwaltschaft beim Landgericht Dresden 1120/2529 Bt 19.

356 Henning to Schwurgericht SHStA Dresden, Staatsanwaltschaft beim Landgericht Dresden 1120/2529 Bt 74.

357 Lempe to Schwurgericht SHStA Dresden, Staatsanwaltschaft beim Landgericht Dresden 1120/2529 Bt79.

358 H Markwardt (2016), 'Der Dresdner 'Euthanasie'-Prozess 1947'. Paper presented at the conference 'Von der Euthanasie zum Holocaust. Parallelität oder Kausalität', Frankfurt am Main, 24-26 November 2016.

359 Record of interview with Nitsche - SHStA Dresden, Staatsanwaltschaft beim Landgericht Dresden 1120/2526 Bt 45.

360 Anklage gegen Werner Heyde - SHStA Dresden, Staatsanwaltschaft beim Landgericht Dresden 1120/2529 Bt 69-72.

361 Lempe to Schwurgericht -SHStA Dresden, Staatsanwaltschaft beim Landgericht Dresden 1120/2532 Bt 109.

362 Report of execution HP Nitsche - SHStA Dresden, Staatsanwaltschaft beim Landgericht Dresden 1120/2533 Bt 46.

363 Strafanträge - SHStA Dresden, Staatsanwaltschaft beim Landgericht Dresden 1120/2534 Bt 35.

PART 2
INTRODUCTION

364 This extract from the Audit Report of the Ministry of the Interior of Wurttemberg, 10 December 1936, is one of a range of records documenting the implementation of the Nazi hereditary health laws included in the DGPPN exhibition, 'Registered, persecuted, annihilated: the sick and disabled under National Socialism' (2014).

365 This extract from a 1940 letter of Pfarrer Ludwig Schlaich, the director of the Protestant mental hospital in Stetten, is one of a number of stories documenting the fears of patient in institutions, also included in the DGPPN exhibition.

366 'Epistemic violence' refers to the process of selectively interpreting data about an 'Other' that depicts them as inferior or probelmatic. Epistemic violence and related key works are discussed further in Chapter 8.

367 P Farmer, et al (2006), 'Structural Violence and Clinical Medicine', *PLoS Medicine*. 3: e449. https://doi.org/10.1371/journal.pmed.0030449

CHAPTER 7
WAS ELVIRA HEMPEL A SURVIVOR OF THE HOLOCAUST?

368 L Alexander (1948), 'War crimes and their motivation: socio-psychological structure of SS and criminalization of society', *Journal of Criminal Law and Criminology*, 39: pp.298-326

369 L Alexander (1949a), *Medical Science under Dictatorship*, Waltham MA: Massachusetts Medical Society.

370 L Alexander (1949b), 'Medical Science under Dictatorship', *New England Journal of Medicine*, 241; pp.39–47

371 L Alexander (1949b), ibid, p.40

372 H Friedlander (1997), op cit; R Lifton (1986), op cit.

373   The Hebrew term 'Shoah' (destruction) was used in some settings during the period of actual persecution.

374   Z Garber and B Zuckerman (2004), 'Double Takes: Thinking and Rethinking Issues of Modern Judaism in Ancient Contexts', *Studies in the Shoah*, vol. 26 Lanham, MD: University Press of America, Inc.

375   D Magilow and L Silverman (2015), *Holocaust Representations in History: An Introduction*, New York: Bloomsbury.

376   Yad Vashem 'The Holocaust: Definition and Preliminary Discussion' http://www.yadvashem.org/yv/en/holocaust/resource_centre/the_holocaust.asp

377   Yad Vashem, ibid.

378   See T Snyder (2010), *Bloodlands: Europe between Hitler and Stalin*, New York: Basic Books and; D Bloxham (2009), *The Final Solution: A Genocide*, New York: Oxford University Press.

379   F Musial (2000), 'Konterrevolutionäre Elemente sind zu erschießen', *Die Brutalisierung des deutsch-sowjetischen Krieges im Sommer 1941*, Propyläen: Berlin/München.

380   E Nolte (1986), 'Die Vergangenheit, die nicht vergehen will', *Frankfurter Allgemeine Zeitung*, June 6, 1986. On the issue of the '*Historikerstreit*', see S. Kattago (2001), *Ambiguous Memory: The Nazi Past and German National Identity*. Westport: Praeger.

381   G Rosenfeld (2015), *Hi Hitler! How the Nazi Past Is Being Normalized in Contemporary Culture*, London: Cambridge University Press.

382   R Cohen (2012), 'The Suffering Olympics', *The New York Times*. 31.01.2012. http://www.nytimes.com/2012/01/31/opinion/the-suffering-olympics.html?_r=1&src=tp&smid=fb-share [Accessed 31 December 2016]

383   H Jacobsen (2017), 'Surviving the Holocaust: 'I didn't allow any hatred to grow. But I don't blame those who did', *Guardian Online* 14 Jan 2017 https://www.theguardian.com/world/2017/jan/14/surviving-holocaust-didnt-allow-hatred-grow-memorial-day [Accessed 15 January 2018]

384   D Lipstadt (1993), *Denying the Holocaust: The Growing Assault on Truth and Memory*, New York: Plume. Lipstadt's book gained considerable publicity when in 2000 she prevailed in a high profile libel suit brought against her and her publisher Penguin in the British Courts by British author David Irving, who she had named as a Holocaust denier.

385 D Lipstadt (2017), 'The Trump Administration's Flirtation With Holocaust Denial', *The Atlantic*, January 30 2017. https://www.theatlantic.com/politics/archive/2017/01/the-trump-administrations-softcore-holocaust-denial/514974/ [Accessed 9 June 2017]

386 Z Bauman (1989), *Modernity and the Holocaust*, Ithaca: Cornell University Press.

387 C Tatz and W Higgins (2016), *The Magnitude of Genocide*, Santa Barbara: Praeger, p.74

388 Mason T (1995), 'Intention and explanation A current controversy about the interpretation of National Socialism', in: *Nazism, Fascism and the Working Class*, J Caplan (ed), Cambridge: Cambridge University Press, pp.212-230

389 See K Bracher (1969), *Die deutsche Diktatur. Entstehung, Struktur, Folgen des Nationalsozialismus*, Koln: Kiepenheuer & Witsch; E Jäckel (1969), Hitler's Weltanschauung: Entwurf einer Herrschaft, Stuttgart: Deutsche Verlags-Anstalt.

390 M Broszat (1981), *The Hitler State: The Foundation and Development of the Internal Structure of the Third Reich*, London: Longman.

391 See Y Bauer (2001), *Rethinking the Holocaust*, London: Yale University Press; C Browning (1992), *The path to genocide: essays on launching the final solution*, Cambridge: Cambridge University Press; K Schleunes (1970), *The Twisted Road to Auschwitz; Nazi Policy Toward German Jews, 1933–1939*, Urbana: University of Illinois Press.

392 H Schmuhl (1987), *Rassenhygiene, Nationalsozialismus, Euthanasie. Von der Verhütung zur Vernichtung 'lebensunwerten Lebens' 1890-1945*, Göttingen: Vandenhoeck & Ruprecht.

393 R. Jütte, et al (2011), *Medizin im Nationalsozialismus. Bilanz und perspektiven der Forschung*, Göttingen: Wallstein, pp.214-255

394 C Tatz (2005), 'The doctorhood of genocide', in: *Genocide and Human Rights: A Philosophical Guide*, J Roth (ed), Hampshire: Palgrave Macmillan, pp.82-94

395 C Tatz and W Higgins (2016), ibid, pp.3-4

396 R Lemkin (1947), 'Genocide as a Crime under International Law', *The American Journal of International Law*, 41: pp.145-151

397 R Lemkin (1944), Chapter ix: "Genocide' a New Term and New Conception for the Destruction of Nations', in: *Axis Rule in Occupied Europe - Laws of Occupation – Analysis of Government – Proposals for Redress*, Washington, D.C: Carnegie Endowment for International Peace.

398 United Nations (1948), *Convention on the Prevention and Punishment of the Crime of Genocide* (A/RES/3/260), UN General Assembly, December 1948. Available online: http://www.un-documents.net/a3r260.htm [Accessed 13 September 2017]

399 A Jones (2010), *Genocide: A Comprehensive Introduction*, New York: Routledge; D Mitchell & S Snyder (2003), 'The Eugenic Atlantic: Race, Disability, and the Making of an International Eugenic Science, 1800–1945', *Disability & Society*, 18: pp.843-864

400 See C Kudlick (2003), Disability History: Why We Need Another Other, *American Historical Review*, 108:763-794

401 N Winfield (1999), 'UN failed Rwanda', *Global Policy Forum*, Dec 16th 1999. https://www.globalpolicy.org/component/content/article/201/39240.html [Accessed 14 January 2017]

402 In December 2017, the United Nations human rights commissioner Zeid Ra'ad al-Hussein said an act of genocide against Rohingya Muslims in Rakhine by state forces in Myanmar 'cannot be ruled out': 'Rohingya crisis: UN rights chief 'cannot rule out genocide' *BBC News* 5th December 2017, http://www.bbc.com/news/world-asia-42234469 [Accessed 18 January 2018]. At the time of writing, United Nations-appointed investigators called on the UN Security Council to refer Myanmar to the International Criminal Court to face genocide charges. Releasing its report into the circumstances surrounding the mass exodus of more than 700,000 Rohingya people from Myanmar, beginning in mid-August 2017, the Independent International Fact-Finding Mission on Myanmar concluded that: 'criminal investigation and prosecution is warranted, focusing on the top Tatmadaw [armed forces] generals, in relation to the three categories of crimes under international law; genocide, crimes against humanity and war crimes': 'Myanmar military leaders must face genocide charges – UN report', public statement, *United Nations*, 27 August 2018. [Accessed online https://news.un.org/en/story/2018/08/1017802]

403 Other ethnic groups within the Ottoman empire, including Greeks and Assyrians, were also subject to genocidal acts in this period. See R Kevorkian (2006), *The Armenian Genocide: A History*, London: IB Tauris. For an account of Turkish denialism, see F Göçek (2015), *Denial of violence: Ottoman past, Turkish present and collective violence against the Armenians, 1789-2009*, New York: Oxford University Press.

404 C Tatz (2017), *Australia's Unthinkable Genocide*, Bloomington Xlibris; J Balint (2017), 'Too near and too far: Australia's reluctance to name and prosecute genocide', in: N Marczak & K Shields (eds), *Genocide Perspectives V- A Global Crime, Australian Voices*, Sydney: UTS ePRESS, pp.51-68

405 M Shaw (2015), *What is Genocide?* Cambridge: Polity.

406 I Charney (Ed) (1999), *Encylopedia of Genocide Vol 1*, Santa Barbara: ABC-Clio, p.14

407 J Armatta (2010), *Twilight of Impunity: The War Crimes Trial of Slobodan Milosevic*, Durham: Duke University Press. In 2016 the ICTY prosecuted a separate case against former Bosnian Serb leader Radovan Karadžić and concluded there was insufficient evidence to find that Slobodan Milosevic agreed with the common plan to create territories ethnically cleansed of non-Serbs during the Bosnian War of 1992 to 1995.

408 F Campbell (2010), *Contours of Ableism: The Production of Disability and Abledness*, Melbourne: Palgrave Macmillan.

409 In Western moral philosophy, the concept of 'the Other' has become a prominent theme since the Holocaust. The foundational work of Simone de Beauvoir on the social construction of the feminine 'Other' and its broad implications, and the work of philosopher Emmanuel Levinas, are profoundly influential. Written in the shadow of the Holocaust and the other violent excesses of the twentieth century, Levinas' work situates the ethical obligation to the 'Other' at the centre of ethics and understands genocide and other atrocities as emerging from humanity's abandonment of this obligation. See: S de Beauvoir (1949), *Le Deuxième sexe*, Paris: Gallimard; E Levinas (1969), *Totality and Infinity: An Essay on Exteriority*, A Lingis (trans), Pittsburgh, PA: Duquesne University Press; E Levinas (1978), *Otherwise than Being or Beyond Essence*, A Lingis (trans), Dordrecht and Boston, MA: Kluwer Academic Publishers.

410 A Maisel (1946), 'Bedlam 1946 - Most Us Mental Hospitals Are a Shame and a Disgrace', *Life Magazine*, May 6, 1946.

411 S Taylor (2009), *Acts of Conscience: World War II, Mental Institutions, and Religious Objectors*, Syracuse, N.Y: Syracuse University Press.

412 H Jones (2013), *Byberry State Hospital*, Charleston: Arcadia.

413 A Deutsch (1948), *The Shame of the States*, New York: Harcourt Brace, p.69

414 United Nations Special Rapporteur on the Right to Health (2017), 'Report of the Special Rapporteur on the right of everyone to the enjoyment of the highest attainable standard of physical and mental health'. A/HRC/35/21, report to the UN Human Rights Council, Thirty-fifth session, 6-23 June 2017.

415 United Nations Special Rapporteur on the rights of persons with disabilities (2017), 'Report of the Special Rapporteur on the rights of persons with disabilities', A/72/133, report to the General Assembly, Seventy-second session, July 2017, page 11

416 Special Rapporteur on the rights of persons with disabilities (2017), ibid.

417 M De Hert et al (2011), 'Physical Illness in Patients with Severe Mental Disorders. I. Prevalence, Impact of Medications and Disparities in Health Care', *World Psychiatry*, 10: pp.52–77; S Kisely et al (2013), 'Reducing All-Cause Mortality among Patients with Psychiatric Disorders: A Population-Based Study', *CMAJ: Canadian Medical Association Journal*, 185 E50-E56; DOI: https://doi.org/10.1503/cmaj.121077.

418 Royal Australian and New Zealand College of Psychiatrists (2016), *The Economic Cost of Serious Mental Illness and Comorbidities in Australia and New Zealand*, Melbourne: RANZCP.

419 C Armitage (2016), 'A Kind of Creeping Euthanasia: Needless Early Deaths among Mentally Ill Cost the Economy $15 Billion a Year', *Sydney Morning Herald*. http://www.smh.com.au/national/health/a-kind-of-creeping-euthanasia-needless-early-deaths-among-mentally-ill-cost-the-economy-15-billion-a-year-20160408-go1nnc.html.

CHAPTER 8
THE ETHICAL DIMENSION

420 L Terry et al (1964), *Smoking and Health: Report of the Advisory Committee to the Surgeon General of the United States*, Washington: U-23 Department of Health, Education, and Welfare, Public Health Service Publication No. 1103.

421 A Brandt (2007), *The Cigarette Century: The Rise, Fall, and Deadly Persistence of the Product that Defined America*, New York: Basic Books.

422 R Proctor (2012), 'The history of the discovery of the cigarette–lung cancer link: evidentiary traditions, corporate denial, global toll', *Tobacco Control*, 21: pp.87-91

423 R Proctor (1999), *The Nazi War on Cancer*, Princeton: Princeton University Press.

424 R Proctor (1999), ibid.

425 T Chelouche & G Brahmer (2013), *Casebook on Bioethics and the Holocaust*, Haifa: UNESCO Chair in Bioethics, pp.11-13

426 *Oxford English Dictionary*, Third Edition.

427 M Godwin (2015), 'Sure, call Trump a Nazi. Just make sure you know what you're talking about', *The Washington Post*. https://www.washingtonpost.com/posteverything/wp/2015/12/14/sure-call-trump-a-nazi-just-make-sure-you-know-what-youre-talking-about/

428 M Burleigh (1997), op cit.

429 S Mercer and R MacDonald (2007),'Disability and human rights', *The Lancet*, 370: pp.548-549

430 I Kerridge & C Stewart (2009), *Ethics and law for the health professions*, Sydney: The Federation Press, p.1

431 Proctor (1999), ibid.

432 While there was opposition to smoking under the Nazis in Germany, it has been argued that there was no uniform or consistent Nazi policy to combat smoking, and anti-smoking policies were not consistently enforced. See E Bachinger, et al (2008), 'Tobacco policies in Nazi Germany: not as simple as it seems', *Public Health*, 122(5), pp.497–505

433 Z Bauman (1989), op cit.

434 M Foucault (1981), *The History of Sexuality. Volume I - an Introduction*, London: Harmondsworth: Penguin. Foucault evolved these ideas later in a series of lectures at the *Collège* de France in Paris in the winter term of 1978. See M Foucault (2004), *The Birth of Biopolitics – Lectures at the Collège de France 1978-1979*, G Burchell (trans), New York: Picador.

435 Foucault considers knowledge is always an exercise of power and power always a function of knowledge. He uses the term 'Power/Knowledge' to convey this fact. See M Foucault (1969), *The Archaeology of Knowledge*, A Sheridan (trans), New York: Pantheon Books.

436 Thomas Foth has framed the actions of nurses in the *Krankenmorde* in terms of biopower. See T Foth (2013), 'Understanding 'Caring' through Biopolitics: The Case of Nurses under the Nazi Regime', *Nursing Philosophy*, 14: pp.284-294

437 M Foucault (2003), *Society Must Be Defended: Lectures at the Collège de France, 1975-1976*, D Macey (trans), New York: Picador. The other important thinker in this area is Italian philosopher Giorgio Agamben who wrote of the concentration camp (both from the Nazi period and its modern equivalents in refugee detention centres) as the ultimate manifestation of biopower, see G Agamben (1988), *Homo Sacer: Sovereign Power and Bare Life*, Stanford: Stanford University Press.

438 P Bourdieu, et al (2000), *Weight of the World: Social Suffering in Contemporary Society*, Stanford, Stanford University Press. Also see D Swartz (1997), *Culture and Power: The Sociology of Pierre Bourdieu*. Chicago, University of Chicago Press.

439 T Teo (2010) 'What is Epistemological Violence in the Empirical Social Sciences?', *Social and Personality Psychology Compass* 4/5: pp.295–303. The Indian scholar Gaytari Spivak cites as a paradigmatic example of epistemic violence; the exercise of colonial power over the colonised through denial or delegitimation of their identity, and links it to Foucault's observation of medicine's epistemic violence against madness in the eighteenth century. See: G Spivak (1988), 'Can the Subaltern Speak?' in *Marxism and the Interpretation of Culture*. C Nelson and L Grossberg (eds), Urbana, University of Illinois Press, pp.271-313

440 M Foucault (1961/2006), *History of Madness*, J Khalfa and J Murphy (trans), New York: Routledge.

441 M Robertson and G Walter (2013), *Ethics and Mental Health: Patient, Profession and Community*, Boca Raton CRC Press.

442 For an interesting formulation of biopower applied to the public health problem of obesity, see C Mayes (2016), *The Biopolitics of Lifestyle*, Oxford: Routledge.

443 M Tinetti (2012), 'The Retreat From Advanced Care Planning', *Journal of the American Medical Association*, 307: pp.915–916

444 S Palin (2009), 'Statement on the Current Health Care Debate', 'Sarah Palin' Facebook account, 9 August 2009: https://www.facebook.com/notes/sarah-palin/statement-on-the-current-health-care-debate/113851103434/ [Accessed 18 September 2017]

445 Larouche PAC (27 December 2010) *Yes, Virginia, There Are Death Panels; Obama To Promote Euthanasia Counselling in End-Run Around Congress* http://archive.larouchepac.com/node/17001

446 The claims were named as PolitiFact's 2009 'Lie of the Year': http://www.politifact.com/truth-o-meter/article/2009/dec/18/politifact-lie-year-death-panels/

447 Tinetti (2102), ibid. Any savings from advanced care planning—in which some patients 'prioritise aggressive treatment focused on prolonging life as long as possible; others want care focused on comfort, function, and quality of life'—come from avoiding care 'unwanted' by patients, not denying needs or preferences.

448 Questions of whether it is ethical to withhold live-saving treatment in circumstances of a prolonged vegetative state do not appear to be informed by eugenic-inspired

mass murder. The 'Cruzon case' involved a 32-year-old woman in a persistent vegetative state after a motor vehicle accident. The case was the first 'right to die' case argued before the US Supreme Court and led to the development of the precedent of 'advanced directives'. The use of the Holocaust as a precedent argument is discussed in A Caplan (1992), op cit.

449 'Testimony Karl Brandt July 19th 1947', *United States v. Karl Brandt et al* (1949), *The Medical Case, Trials of War Criminals before the Nuremberg Military Tribunals under Control Council Law No. 10* Washington: U.S. Government Printing Office.

450 Karl Brandt interview with Leo Alexander 26 November 1946. *United States of America-v-Karl Brandt et al.* (Case I), Nov. 21, 1946 - Aug. 20, 1947. Microfilm Publication M887 (Medical Case), NARA - Records of the U. S. Nuremberg War Crimes Trials.

451 E Burdett (2011), '*The Continent of Murder: Disability and the Nazi 'Euthanasia' Programme in the Euthanasia Debates of Britain and the United States, 1945-Present*', PhD thesis: University College London.

452 Pfeifer to Schwurgericht Dresden - SHStA Dresden, Staatsanwaltschaft beim Landgericht Dresden 1120/2531 Bt 136-40.

453 See B Mishara & D Weisstub (2013), 'Premises and evidence in the rhetoric of assisted suicide and euthanasia', *Int J Law Psychiatry*, 36: pp.427-435

454 P Singer (1993), *Practical Ethics*, New York: Cambridge University Press, p.200

455 See J Rachels (1996), *The End of Life*, Oxford: Oxford University Press, pp.175–178

456 'A to Z' not 'A to B' is a deliberate device to highlight that there are numerous intermediary steps and not an inevitable progression from a well-intentioned act to a situation of calamity or malevolence. See D Enoch (2001), Once You Start Using Slippery Slope Arguments You're On a Very Slippery Slope, *Oxford Journal of Legal Studies*, 21: pp.629-647

457 E Emanuel (1994), 'The history of euthanasia debates in the United States and Britain', *Annals of Internal Medicine*, 121: pp.793-802

458 See J Keown (2002), *Euthanasia, Ethics and Public Policy: An Argument Against Legalisation*, London: Cambridge University Press.

459 Emanuel also sees that analogous historical circumstances to political and cultural extremism in the 1930s 'suggests an association between interest in legalizing euthanasia and moments when Social Darwinism and raw individualism, free markets

and wealth accumulation, and limited government are celebrated.' See E Emanuel (1998) 'Why Now?' in: E Emanuel, *Regulating how We Die: The Ethical, Medical, and Legal Issues Surrounding Physician-assisted Suicide*, Cambridge MA: Harvard University Press, pp.175-202

460 See P Singer (1993), ibid, p.345

461 See P Singer et al (1995), 'Double Jeopardy and the Use of QALYS in Health Care Allocation', *Journal of Medical Ethics*, 21; and P Singer (1993), op cit.

462 P Singer (2017), 'We should end the suffering of patients who know they are dying and want to do so peacefully', *Guardian Australia* https://www.theguardian.com/commentisfree/2017/sep/18/we-should-end-the-suffering-of-patients-who-know-they-are-dying-and-want-to-do-so-peacefully [Accessed 20 September 2017]

463 Dyer, O, et al (2015), 'Assisted dying: law and practice around the world', *British Medical Journal Online*, 2015; 351 DOI: https://doi.org/10.1136/bmj.h4481

464 A number of attempts to legalise euthanasia in these and other Australian jurisdictions have been made in recent decades. The only successful bill was the Northern Territory's *Rights of the Terminally Ill Act*, which was overturned by the national parliament in 1997, less than a year after it had commenced operation.

465 E Emanuel, et al (2016), 'Attitudes and Practices of Euthanasia and Physician-Assisted Suicide in the United States, Canada, and Europe', *Journal of the American Medical Association*, 316(1): pp.79-90; E Emanuel (2017), 'Euthanasia and physician-assisted suicide: focus on the data', *Medical Journal of Australia*, 206: pp.339-340

466 K Chambaere, et al (2011), 'Palliative Care Development in Countries with a Euthanasia Law – Report for the Commission on Assisted Dying Briefing Papers', October 4 2011. Available at http://www.rpcu.qc.ca/pdf/documents/PalliativeCareDevInCWAEuthanasiaLaw.pdf; J Bernheim, et al (2014), 'State of Palliative Care Development in European Countries with and without Legally Regulated Physician-Assisted Dying', *Health Care*, 2: pp.10-14

467 Commission Late Pregnancy Termination and Termination of Life in Newborns 2016, Annual Reports, http://www.lzalp.nl/over-de-commissie/jaarverslagen. Also see, See N Francis (2016), 'Neonatal deaths under Dutch Groningen Protocol very rare despite misinformation contagion', *Journal of Assisted Dying*, 1: pp.7–19

468 S Dierickx, et al (2017), 'Euthanasia for people with psychiatric disorders or dementia in Belgium: analysis of officially reported cases', *BMC Psychiatry*, 17: p.203

469 L Thienpont, et al (2015), 'Euthanasia requests, procedures and outcomes for 100 Belgian patients suffering from psychiatric disorders: a retrospective, descriptive study', *British Medical Journal Open*, 5(7): e007454.

470 S Kim et al (2016), 'Euthanasia and Assisted Suicide of Patients With Psychiatric Disorders in the Netherlands 2011 to 2014', *Journal of the American Medical Association-Psychiatry*, 73: pp.362–368

471 'Netherlands may extend assisted dying to those who feel 'life is complete''*The Guardian Online*, https://www.theguardian.com/world/2016/oct/13/netherlands-may-allow-assisted-dying-for-those-who-feel-life-is-complete [Accessed 16 January 2018]

472 P Appelbaum (2016), 'Physician-Assisted Death for Patients With Mental Disorders—Reasons for Concern', *Journal of the American Medical Association – Psychiatry*, 73: pp.325–326

473 World Medical Association 'Declaration on Euthanasia' (1987, reaffirmed 2015) https://www.wma.net/policies-post/wma-declaration-on-euthanasia/ [Accessed 20 September 2017]

474 'World Medical Association reiterates strong opposition to physician assisted suicide and to Australian bill' (27 October 2017) https://www.wma.net/news-post/world-medical-association-reiterates-strong-opposition-to-physician-assisted-suicide-and-to-australian-bill/

475 On this point, the involvement of US health insurers in such decision-making, particularly following the public endorsement of PAS in California in late 2016, presents a likely ethical challenge around whether PAS is to be categorized as a 'clinical intervention' and what cost imperatives may emerge from this. See S Stefan (2016), *Rational Suicide, Irrational Laws: Examining Current Approaches to Suicide in Policy and Law*, New York: Oxford University Press.

476 Gannon, M (2017), 'Euthanasia And Marriage Equality Are Life-Changing Issues That Demand Our Attention', *Huffington Post Online* http://www.huffingtonpost.com.au/michael-gannon/euthanasia-and-marriage-equality-are-two-life-changing-issues-that-demand-our-attention_a_23214217/?utm_hp_ref=au-homepage

477 Carr lives with arthrogryposis multiplex congenital, an inherited condition characterised by multiple joint contractures.

478 E Carr (2016), 'Legalising assisted dying is dangerous for disabled people. Not compassionate'. https://www.theguardian.com/commentisfree/2016/sep/09/legalising-assisted-dying-dangerous-for-disabled-not-compassionate.

479 See J Stramondo (2016), 'Why Bioethics Needs a Disability Moral Psychology,' *Hastings Centre Report*, 46: pp.22-30

480 S Young 'Disability – a fate worse than death?', *Guardian Australia Online* https://www.theguardian.com/commentisfree/2013/oct/18/disability-euthanasia-assisted-dying [Accessed 21 September 2017]

481 The grassroots organization 'Not Dead Yet' has emerged as a voice against this sentiment towards euthanasia of the disabled. See 'Not Dead Yet' http://notdeadyet.org/

482 For a detailed consideration of this history, see D Braddock and S Parish (2002), 'An Institutional History of Disability', in: D Braddock (ed) *Disability at the Dawn of the 21st Century and The State of the States*, Washington: American Association Intellectual Disabilities, pp.11-68

483 C Poore (2007), *Disability in Twentieth Century German Culture*, Ann Arbor: University of Michigan Press.

484 For a detailed discussion of the history of the disability rights movement in Germany, see S Köbsell (2006), 'Towards Self-Determination and Equalization: A Short History of the German Disability Rights Movement', *Disability Studies Quarterly*, 26(2).

485 See U Sierck &N Radtke (1984), *Die Wohltätermafia. Vom Erbgesundheitsgericht zur Humangenetischen Beratung*, Hamburg: Sierck Verlag; and T Degener & S. Köbsell (1992), *Hauptsache, es ist gesund'?! Weibliche Selbstbestimmung unter humangenetischer Kontrolle*, Hamburg: Konkret Literatur Verlag.

486 L Sherbin and J Kennedy (2017), *Disabilities and Inclusion*, New York: Centre for Talent Innovation http://www.talentinnovation.org/_private/assets/DisabilitiesInclusion_KeyFindings-CTI.pdf

487 P Handley (2003), 'Theorizing Disability: Beyond Common Sense', *Politics*, 23: pp.109-118

488 M Oliver (1998), 'Theories in Health Care and Research: Theories of Disability in Health Practice and Research', *British Medical Journal*, 317: pp.1448 -1449

489 Supreme Court of the United States (1927), 'Carrie Buck v. John Hendren Bell, Superintendent of State Colony for Epileptics and Feeble Minded', 274 U.S. 200 (more) 47 S. Ct. 584; 71 L. Ed. 1000; 1927 U.S. LEXIS 20. This was discussed in Chapter 2.

490 P Risher and S Amorosi (1998), 'The 1998 National Organization of Disabilities / Harris Survey of Americans with Disabilities', New York: Harris (Louis) and Associates, Inc http://eric.ed.gov/?id=ED422692.

491 S Levi (2006), 'Ableism', in: *Encyclopedia of Disability*, GL Albrecht (ed), Thousand Oaks: SAGE Publications, pp.1-5. Levi cites P Terry (1996), 'Preparing Educational Leaders to Eradicate the 'isms.", Paper presented at the *Annual International Congress on Challenges to Education: Balancing Unity and Diversity in a Changing World*, Palm Beach, Aruba, July 10–12, Obtained from Educational Resources Information Center (ERIC).

492 Levi (2006), ibid.

493 R Bourne (1911), *The Handicapped - by One of Them*, Online from the Disability History Museum http://www.disabilitymuseum.org/dhm/lib/detail.html?id=2009&page=all [Accessed 18 January 2018]

494 Levi (2006), op.cit.

495 Levi (2006), op cit.

496 United Nations Special Rapporteur on the rights of persons with disabilities (2017). *Report of the Special Rapporteur on the rights of persons with disabilities*, A/72/133, report to the General Assembly, Seventy-second session, July 2017.

497 World Health Organisation & World Bank (2011), *World Report on Disability*, Geneva: WHO, p xxi.

498 World Health Organisation & World Bank (2011), ibid, p.262

499 World Health Organisation & World Bank (2011), op cit, p.263

500 World Health Organisation & World Bank (2011), op cit, p.263

501 World Health Organisation & World Bank (2011), op cit, pp.2-3

502 https://www.un.org/development/desa/disabilities/convention-on-the-rights-of-persons-with-disabilities.html

503 R Kayess, P French (2008), 'Out of Darkness into Light? Introducing the Convention on the Rights of Persons with Disabilities', *Human Rights Law Review* 8: pp.1–34

504 United Nations Division for Social Policy and Development Disability. The Convention in Brief. 2006; https://www.un.org/development/desa/disabilities/convention-on-the-rights-of-persons-with-disabilities/the-convention-in-brief.html.

505 Union of the Physically Impaired Against Segregation and The Disability Alliance (1976), *Fundamental Principles of Disability*, London: UPIAS pp.3-4. The UK-based Union of the Physically Impaired Against Segregation met the Disability Alliance in 1975 to consider ways in which disabled people could become more active in the disability field, agreeing to Union-proposed principles: 'disability is a situation, caused by social conditions, which requires for its elimination, (a) that no one aspect such as incomes, mobility or institutions is treated in isolation, (b) that disabled people should, with the advice and help of others, assume control over their own lives, and (c) that professionals, experts and others who seek to help must be committed to promoting such control by disabled people.' The *Fundamental Principles Of Disability* provides a summary of the meeting discussion and each organisation's commentary on it. Copies of it and other related UPIAS documents can be found in University of Leeds Centre for Disability Studies archive: https://disability-studies.leeds.ac.uk/library/author/upias/

506 M Oliver (1996), *Understanding disability: from theory to practice*, Basingstoke: Macmillan.

507 M Oliver (1983) *Social Work with Disabled People*, Basingstoke: Macmillan; and M Oliver (1990), *The Politics of Disablement*, Basingstoke: Macmillan.

508 L Staniland (2012), *Public Perceptions of Disabled People: Evidence from the British Social Attitudes Survey 2009*, London: Office for Disability Issues.

509 S Roush (1986), 'Health professionals as contributors to attitudes toward persons with disabilities: a special communication', *Physical Therapy*, 66: pp.1551–1554; and R Northway (1997), 'Disability and oppression: some implications for nurses and nursing', *Journal of Advanced Nursing*, 26: pp.736–743

510 A Chappel (1992), 'Towards a sociological critique of the normalization principle', *Disability Handicap and Society*, 7: pp.1-2

511 Australian disability advocate Stella Young described this phenomenon as the emotional response to 'an image of a person with a disability, often a kid, doing something completely ordinary', http://www.abc.net.au/news/2012-07-03/young-inspiration-porn/4107006

512 For other contemporary accounts of disability, bioethics and biopolitics, see: JL Scully (2008), *Disability bioethics: Moral bodies, Moral difference*, Gloucester: Rowman

& Littlefield Publishers; T Shakespeare (2006), *Disability rights and wrongs*, London: Routledge; D Mitchell & S Snyder (2015), *The Biopolitics of Disability: Neoliberalism, Ablenationalism, and Peripheral Embodiment*, Ann Arbor: University of Michigan Press.

513   M Oliver (2013), 'The Social Model of Disability: Thirty Years On', *Disability & Society*, 28: pp.1024-1026

514   See D Harvey (2005), *A Brief History of Neoliberalism*, London: Oxford University Press.

515   See R Owen & S Parker Harris (2012), 'No rights without responsibilities: Disability rights and neoliberal reform under New Labour', *Disability Studies Quarterly*, 32. http://dx.doi.org/10.18061/dsq.v32i3

516   K Soldatic & A Chapman (2010), 'Surviving the Assault? The Australian Disability Movement and the Neoliberal Workfare State', *Social Movement Studies*, 9: pp.139-154

517   GK Chesterton (1922), *Eugenics and Other Evils*, London: Cassell and Company Limited.

518   World Health Organization (2014), 'Gender and Reproductive Rights'. http://www.who.int/reproductivehealth/en/; also see E Dyck (2013), *Facing Eugenics: Reproduction, Sterilization, and the Politics of Choice*, Toronto: University of Toronto Press.

519   S Goering (2014), 'Eugenics', *The Stanford Encyclopedia of Philosophy* (Fall 2014 Edition), Edward N. Zalta (ed.), URL: https://plato.stanford.edu/archives/fall2014/entries/eugenics/

520   See C Saleeby (1909), *Parenthood and race culture: An outline of eugenics*, New York: Moffat, Yard, and Co. Saleeby was influential in the eugenic movement in the UK and USA and successfully advocated the creation of a Ministry of Health in Great Britain in 1919.

521   S Goering (2014), ibid.

522   See D Plotz (2005), *The Genius Factory: The Curious History of the Nobel Prize Sperm Bank*, New York: Random House. Graham's sperm bank closed two years after his death in 1997.

523   R Hernstein & C Murray (1994), *The Bell Curve: Intelligence and Class Structure in American Life*, New York: The Free Press.

524   J Bickman (2012), *Should addicts be sterilized?* Salon.com, Wednesday, May 2, 2012. http://www.salon.com/2012/05/02/should_addicts_be_sterilized_salpart/ [Accessed 28 March 2017]

525   U.S. Department of Energy, Office of Science, & Office of Biological and Environmental Research (2013), Human Genome Project. *Human Genome Project Information Archive*, Retrieved from http://www.ornl.gov/hgmis. [28 March 2017]

526   Prenatal diagnosis was developed throughout the 20th century and the first reported prenatal diagnosis for sickle cell anemia using molecular genetic techniques was reported in 1987. There are now prenatal tests available to diagnose hundreds of genetic conditions. See E Parens & A Asch (eds) (2000), *Prenatal Testing and Disability Rights*, Georgetown: Hastings Center Studies in Ethics series. In the process of IVF, preimplantation genetic diagnosis (PGD) and preimplantation genetic screening (PGS) refers to testing embryos for specific genes (such as those associated with cystic fibrosis, Fragile X syndrome, Huntington's disease), possession of the BRCA1/BRCA2 gene (causing significantly increased risk of breast or ovarian cancers) or to detect chromosomal conditions (such as Down syndrome).

527   D Cyranoski (2017), 'China's embrace of embryo selection raises thorny questions', *Nature* 548: pp.272-274

528   See A Asch (2000), 'Why I haven't changed my mind about prenatal diagnosis: Reflections and reminders', and M Saxton (2000), 'Why members of the disability community oppose prenatal diagnoses and selective abortion', both in E Parens and A Asch, ibid, pp. 147–164; and J Hampton (2005), 'Family eugenics', *Disability and Society*, 20: pp.553–561

529   A Cohen (2017), 'Is there such a thing as good eugenics?', *Los Angeles Times* http://www.latimes.com/opinion/op-ed/la-oe-cohen-good-eugenics-20170317-story.html [Accessed 28 March 2017]

530   D Wertz (2002), 'Did eugenics ever die?', *Nature*, 3:408.

531   S Camporesi & G Cavaliere (2018), 'Eugenics and enhancement in contemporary genomics', in S Gibbon, B Prainsack, S Hilgartner, and J Lamoreaux (eds), *Routledge Handbook of Genomics, Health and Society*, London: Routledge. The authors note the first use of the term 'reprogenetics' by geneticist Lee Silver in his 1997 book *Remaking Eden*, New York: Avon Books.

532   S Camporesi & G Cavaliere (2018), ibid.

533   Nuffield Council on Bioethics (2018), *Genome editing and human reproduction: social and ethical issues*, July 2018, London: Nuffield Council on Bioethics. The Nuffield

Council determined 'the potential use of heritable genome editing interventions to influence the characteristics of future generations could be ethically acceptable in some circumstances, so long as it is intended to secure, and is consistent with, the welfare of a person who may be born as a consequence of interventions using genome edited cells; and it is consistent with social justice and solidarity, i.e. it should not be expected to increase disadvantage, discrimination, or division in society'.

534 S Camporesi & G Cavaliere (2018), op cit; X Zeng, et al (2016), 'Localizing NIPT: Practices and meanings of non-invasive prenatal testing in China, Italy, Brazil and the UK', *Ethics, Medicine and Public Health*, 2: pp.392—401; and M Minear, et al (2015), 'Global perspectives on clinical adoption of NIPT', *Prenatal Diagnosis*, 35: pp.959–967

535 M Minear et al (2015), ibid, p966

536 Minear et al (2015), op cit; J Dickinson and M Gannon (2017), 'Non-Invasive Prenatal Testing – Ethical & Medico-legal Issues', 'Defence update - 15 November 2017' *MDA National*: https://defenceupdate.mdanational.com.au/Articles/nipt-ethical-issues

537 ResearchAndMarkets.com (2018), 'Non-Invasive Prenatal Testing (NIPT): Global Market Opportunity Analysis and Forecast 2018-2025 - China NIPT Market Expected to Cross $400 Million by 2025, Boosted by Two-Child Policy', news release [includes link to report] 29 June 2018: https://globenewswire.com/news-release/2018/06/29/1531566/0/en/Non-Invasive-Prenatal-Testing-NIPT-Global-Market-Opportunity-Analysis-and-Forecast-2018-2025-China-NIPT-Market-Expected-to-Cross-400-Million-by-2025-Boosted-by-Two-Child-Policy.html

538 ResearchAndMarkets.com (2018), ibid.

539 J Dickinson and M Gannon (2017), ibid.

540 C Mills (2017), 'Biopolitics and human reproduction', in: S. Prozorov and S Rentea (eds), *The Routledge Handbook of Biopolitics*, Abingdon: Routledge, pp.281-294

541 For a comprehensive account of the interaction of biopower and neoliberalism, see P Miller and N Rose (2008), *Governing the Present: Administering Economic, Social and Personal Life*, Cambridge: Polity; and L Mavelli (2016), 'Governing the resilience of neoliberalism through biopolitics', *European Journal of International Relations*, 23: pp.489–512

542 Nuffield Council on Bioethics (2017), 'New pregnancy testing technique needs

limits says ethics body' http://nuffieldbioethics.org/news/2017/pregnancy-testing-technique-limits-ethics-body.

543   L Burch (2017), 'A world without Down's syndrome? Online resistance on Twitter: #worldwithoutdowns and #justaboutcoping', *Disability & Society* 32:7, pp.1085-1089. Published online: 25 May 2017.

544   X Zeng, L Zannoni, I Löwyc, S Camporesi (2016), 'Localizing NIPT: Practices and meanings of non-invasive prenatal testing in China, Italy, Brazil and the UK', *Ethics, Medicine and Public Health*, 2: pp.392-401

545   A Newson (2016), 'Why information and choice won't solve all of NIPT's ethical problems', *BioNews*, 30:866 https://www.bionews.org.uk/page_95667.

546   C Mills (2015), 'Liberal Eugenics, Human Enhancement and the Concept of the Normal', in D Meacham (ed) *Medicine and Society, New Perspectives in Continental Philosophy*, Dordrecht: Springer, pp.179–194

547   J Lederberg (1963), 'Molecular biology, eugenics and euphenics', Nature, 198: pp.428–429; and J Lederberg (1966), 'Experimental genetics and human evolution', *The American Naturalist*, 100: pp.519–531

548   J Huxley (1957), 'Transhumanism', in: *New Bottles for New Wine*, London: Chatto & Windus, pp.13-17

549   F Nietzsche (1896/2005), *Thus Spoke Zarathustra*, G Parkes (Trans), Oxford: Oxford World's Classics.

550   Bauman (1989), op cit.

551   H Czech (2014), 'Der Spiegelgrund-Komplex: Kinderheilkunde, Heilpädagogik, Psychiatrie und Jugendfürsorge im Nationalsozialismus', *Österreichische Zeitschrift für Geschichtswissenschaften*, 25: pp.189-214

552   'Biopsychiatry, Genetic Engineering, and 'Euthanasia' Today', in the exhibition 'War Against the 'Inferior': On the history of Nazi Medicine in Vienna', Pavilion V, Otto Wagner-Spital, Vienna, Austria (accessible online http://gedenkstaettesteinhof.at/en/exibition/18-biopsychiatry-genetic-engineering-and-euthanasia-today).

CHAPTER 9
REMBRANCE, COMMEMORATION, MEMORIALISATION

553 Patient File Anstalt Eglfing Haar *Archiv des Bezirk Oberbayern*.

554 See M Kater (1989), op cit.

555 The 'Sewering affair' was brought to light by Canadian academic William Seidelman, see W Seidelman (2014), "Requiescat sine Pace': Recollections and Reflections on the World Medical Association, the Case of Prof. Dr Hans Joachim Sewering and the Murder of Babette Fröwis' in V. Roelcke , S. Topp, E Lepicard (eds), *Silence, Scapegoats, Self-Reflection - The Shadow of Nazi Medicine and Bioethics*, Gottingen, V&R Unipress: pp.281–300

556 Seidelman (2014), ibid.

557 Seidelman (2014), op cit. p.290

558 See 'Professor Sewering's resignation from the WMA President-Elect' (1993), *World Medical Journal*, 39: pp.22-23

559 W Seidelman (1996), 'Vergangenheit: One cannot ignore . . .', *Deutsches Arzteblat*, 93: A-1082 / B-902 / C-844.

560 H de Quetteville (2008), 'German doctor 'who sent 900 children to Nazi camp' honoured', *Daily Telegraph* http://www.telegraph.co.uk/news/worldnews/2027938/German-doctor-who-sent-900-children-to-Nazi-camp-honoured.html

561 *Deutsches Ärtzblatt* (2010), 107: pp.28-29

562 The Nuremberg Declaration of the 2012 German Medical Assembly. http://www.bundesaerztekammer.de/fileadmin/user_upload/downloads/Nuremberg_Declaration_EN.pdf

563 R Shmuel (2012), 'Reflections on the Nuremberg declaration of the German medical assembly', *Israel Medical Association Journal*, 14:532-534

564 At its first general assembly meeting in 1947, the WMA adopted a statement 'War Crimes and Medicine: The German Betrayal and a Restatement of the Ethics of Medicine'. The WMA condemned the crimes of German doctors and endorsed their punishment. The inaugural edition of the *WMA Bulletin*, released in 1949, published

the 1947 statement along with 'The Dedication of the Physician'. The WMA expressed its 'astonishment' that no sign had come from Germany that doctors were ashamed of their role in the crimes of National Socialism. See S Perley (1992), 'The Nuremberg Code: An International Overview', in G Annas and M Grodin (eds) op cit, pp.149–173; and MH Kater (1997), 'The Sewering Scandal of 1993 and the German Medical Establishment', in M Berg and G Cocks (eds), *Medicine and Modernity: Public Health and Medical Care in Nineteenth - and Twentieth-Century Germany*, Washington DC: German Historical Institute, pp.213-234. The WMA was also criticised for its refusal in the late 1970s to exclude the South African Medical Association from its ranks in the aftermath of the death of political activist Steve Biko.

565   A Mitscherlich and M. Mitscherlich (1975), *The Inability to Mourn. Principles of Collective Behavior*, New York : Grove Press.

566   N Frei (2002), *Adenauer's Germany and the Nazi Past: The Politics of Amnesty and Integration*, New York: Columbia University Press.

567   H Ehrlich (1962), 'The Swastika Epidemic of 1959-1960: Anti-Semitism and Community Characteristics', *Social Problems*, 9: pp.264–272

568   T Adorno (2005), 'The Meaning of Working Through the Past', *Critical Models: Interventions and Catchwords*, H. Pickford (trans), New York: Columbia University Press, pp.89-103

569   F Fischer (1986), *From Kaiserreich to Third Reich: Elements of Continuity in German History 1871-1945*, London: Allen & Unwin.

570   See H Wehler (1981), 'Deutscher Sonderweg 'oder allgemeine Probleme des westlichen Kapitalismus', *Merkur*, 5: pp.478–487

571   A Taylor (1945), *The Course of German History*, London: Hamish Hamilton. It is interesting to note that in 2016 German Chancellor, Angela Merkel, was blamed for having lead Germany on a '*Sonderweg*' when in 2015, contrary to previous traditions in German politics and European Union policy, she opened the country's borders allowing almost one million of refugees to enter Germany. See J Kocka (2016), 'Ein neuer deutscher Sonderweg', in: *Neue Gesellschaft Frankfurter Hefte*, 3, pp.14-17

572   For a specific consideration, see P Baldwin (1990), *Reworking the Past: Hitler, the Holocaust and the Historians Debate*, Boston: Beacon Press. Nolte's position is articulated in E Nolte (1987), *Das Vergehen der Vergangenheit: Antwort an meine Kritiker im sogenannten Historikerstreit*, Berlin: Ullstein. Habermas's response to Nolte appeared in J Habermas (1987), *Eine Art Schadensabwicklung: kleine politische Schriften VI*, Frankfurt a.M: Suhrkamp.

573 D Goldhagen (1996), *Hitler's Willing Executioners: Ordinary Germans and the Holocaust*, New York: Knopf.

574 Eichmann had escaped Allied captivity and lived in Buenos Aires under the name Ricardo Klement. Mossad agents abducted him in 1960 as he walked home and smuggled him to Israel where he faced trial for crimes against the Jewish people. He was convicted and hanged in Jerusalem in June 1962.

575 I Bierer (2017), *Albert Speer in the Federal Republic: Dealing with the German Past*, Nuremberg Documentation Centre Party Rally Grounds. For more critical accounts of Speer's complicity with the crimes of the Nazi regime, see J Fest (2001), *Speer: the final verdict*, E Osers & A Dring (trans), New York: Harcourt; and D Van der Vat (1998), *The Good Nazi: The Life and Lies of Albert Speer*, Boston: Houghton Mifflin.

576 This was most evident in the success of Joachim Fest's biography of Hitler. See J Fest (1973), *Hitler: Eine Biographie*, Hamburg Spiegel-Verlag.

577 D Magilow & L Silverman (2015), *Holocaust Representations in History: An Introduction*. New York, Bloomsbury.

578 N Kulish and S Mekhennet (2014), *The Eternal Nazi: From Mauthausen to Cairo, the Relentless Pursuit of SS Doctor Aribert Heim*, New York, Vintage, p.147

579 M Wildt, et al (2004), *Wehrmachtsausstellung* (Wehrmacht Exhibition). Hamburg Institute for Social Research. Hamburg, Hamburger Edition HIS Verlagsges GmbH.

580 The first work on Wehrmacht crimes in the east was: C Streit (1979), *Keine Kameraden. Die Wehrmacht und die sowjetischen Kriegsgefangenen 1941-1945*, Bonn: Dietz.

581 This is a slang term for German soldiers like 'GI' (US Army), 'Poulis' (French Army) 'Tommy' (British Army) or 'Digger' (Australian Army).

582 The controversy prompted by the Wehrmacht exhibition was documented in the 2007 film directed by Michael Verhoeven *Der unbekannte Soldat* (The Unknown Soldier), http://www.imdb.com/title/tt0844462/

583 For a detailed consideration of the notion of a 'mainstreaming' of the Nazi period, see G Rosenfeld (2015), *Hi Hitler! How the Nazi Past Is Being Normalized in Contemporary Culture*, London: Cambridge University Press.

584 The trauma of the German Democratic Republic (GDR) and the human rights abuses of the GDR's secret police (Stasi) has been the basis of another strand of

Erinnerungskultur, evident in productions such as *Goodbye Lenin* (2003), *The Lives of Others* (2006), *Deutschland 1982* (2016) and *The Same Sky* (2017).

585 E Claasen (1987[1969]), *Ich, die Steri*, Bonn: Psychiatrie-Verlag; A Kaufmann (2007 [1986]), *Totenwagen. Kindheit am Spiegelgrund*, Mechthild Podzeit-Lütjen (ed), Wien: Mandelbaum; E Manthey (1994), op cit.

586 R Dome (2008), *Nebel im August*, Munich: Random House.

587 I Buruma (1994), *The wages of guilt - memories of war in Germany and Japan*, New York: New York Review of Books.

588 An Israeli satirist has launched an art project juxtaposing 'selfies' (harvested from social media) of young people behaving in such a way at the Berlin Holocaust memorial, interspersed with archive footage from concentration camps – he has titled the project 'Yolocaust'. https://www.theguardian.com/world/2017/jan/19/yolocaust-artist-shahak-shapira-provokes-debate-over-commemorating-germanys-past [Accessed 20 December 2017]

589 'Auschwitz Selfie Teen Breanna Mitchell Defends Picture: 'I Don't Think I Would Do Anything Different' http://www.usmagazine.com/celebrity-news/news/auschwitz-selfie-teen-breanna-mitchell-defends-picture-video-2014227 [Accessed 3 October 2016]

590 A Bryman (2004), *The Disneyization of Society*, Thousand Oaks: Sage Publications. Here can be seen evidence of the influence of neoliberal commodification, as discussed in Chapter 7.

591 A Alba (2015), *The Holocaust Memorial Museum - Sacred Secular Space*, Basingstoke, Palgrave.

592 Personal communication (EL) with A Thomas, International Coordinator, Stolpersteine (January 2017). For information about the project, see http://www.stolpersteine.eu/en/home/

593 S Knittel (2010), 'Remembering Euthanasia: Grafeneck in the Past, Present and Future', in: *Memorialization in Germany since 1945*, B. Niven and C. Paver (eds), Houndmills: Palgrave, pp.124-133. See also www.dasdenkmaldergrauenbusse.de

594 For a detailed account of these memorial sites in Europe, see G Schmid-Krebs and R. Brüggemann (2007), *Verortungen der Seele - Psychiatrie-Museen und verwandte Einrichtungen in Europa*, Goppingen: Mabuse-Verlag. See also E Light, M Robertson, H Markwardt, C Hanzig, G Walter, W Lipworth (in press), 'Representation and memorialisation of the victims of the Krankenmorde (Nazi

'euthanasia')', in J Barrett, A Alba and D Moses (eds), *The Holocaust, Human Rights and the Contemporary Museum*, Philadelphia: University Pennsylvania Press.

595   This is discussed in some detail in R Surmannn (2014), 'Rehabilitation and Indemnification for the Victims of Forced Sterilization and 'Euthanasia'. The West German Policies of Compensation (*Wiedergutmachung*)', in: V Roelcke, S Topp and E Lepicard (eds), op cit, pp.113-130. See also C Pross (1998), *Paying for the Past - The Struggle over Reparations for Surviving Victims of the Nazi Terror*, Baltimore, Johns Hopkins University Press.

596   See J Peiffer (2006), 'Phases in the Postwar German Reception of the 'Euthanasia Program' (1939–1945) Involving the Killing of the Mentally disabled and its Exploitation by Neuroscientists', *Journal of the History of the Neurosciences*, 15: pp.210-244

597   Hoven was an SS doctor who was initially involved in *Aktion* T4. He is better known as the chief doctor at the Buchenwald concentration camp where his unconsented experiments on prisoners was the main focus of his prosecution.

598   These included former staff from Hadamar, Eichberg and Kalmenhof institutions.

599   Bauer had made a critical contribution to the capture of Adolf Eichmann by Israeli agents in Argentina in 1962 and is also remembered for his prosecution in Frankfurt (1963-65) of Auschwitz perpetrators.

600   See M Burleigh (2002), op cit.

601   R Surmann (2005), Was ist typisches NS-Unrecht? Die verweigerte Entschädigung für Zwangssterilisierte und 'Euthanasie'-Geschädigte, in: M Hamm (ed) *Lebensunwert—zerstörte Leben. Zwangssterilisation und Euthanasie*, Frankfurt M:VAS-Verlag, pp198–211

602   Dietrich Bonhoeffer (son of psychiatrist Karl) was a Lutheran pastor known for his staunch resistance to the Nazi regime, including prominent opposition to 'euthanasia' and the persecution of Jews. He was executed in 1945. Hans and Sophie Scholl, often referred to as 'the Scholl siblings', were members of the White Rose movement, a Munich-based student group active in non-violent resistance to the Nazi regime. Their main act of opposition was distributing flyers around Munich protesting against many of the excesses of the dictatorship. See Chapter 2.

603   *Deutscher Bundestag* 'Remembrance for the victims of National Socialism' https://www.bundestag.de/en/kw04-remembrance/490686 [Accessed 29 January 2017]

Endnotes

604 The GDNP - *Gesellschaft Deutscher Neurologen und Psychiater* re-formed urgently after the war. In the German Democratic Republic (East Germany) the '*Gesellschaft für Psychiatrie und Nervenheilkunde in der DDR*' (Society for Psychiatry and Neurology in the German Democratic Republic) remained the East German Psychiatric profession's main body until re-unification in 1991.

605 For the full text of Schneider's 2010 apology, see DGPPN (2010), 'Psychiatry under National Socialism – Remembrance and Responsibility', https://www.dgppn.de/history/psychiatry-under-national-socialism/speech-professor-schneider.html

606 Dorothea Buck-Zerchin was born in 1917 and became ill with a psychotic episode in 1936. Whilst hospitalised at the Bodelschwingh Institution in Bethel she was sterilized under the Hereditary Health Law in 1937. She was hospitalized many times during and after the war and was a witness to the murders of fellow patients. She learnt pottery and sculpture and worked as both an artist and teacher in Hamburg. Apart from her talents as an artist, she has been a strong life-long advocate of self-help movements for people living with mental illness. Her life and work were honoured with national awards in 1997 and in 2008.

EPILOGUE

607 D Roer (1992), 'Psychiatrie in Deutschland 1933-1945: Ihr Beitrag zur 'Endlösung der sozialen Frage' am Beispiel der Heilanstalt Uchtspringe', *Psychologie und Gesellschaftskritik*, 16: pp.15-37

608 We have no way of establishing what Elvira-Lotte was told of Lisa's death. Elvira was only able to discover her sister Lisa's actual fate when the truth about the 'euthanasia' program emerged after the war.

609 P Kuwert and H Freyberger (2007), 'The unspoken secret: Sexual violence in World War II', *International Psychogeriatrics*, 19: pp.782–784

610 Plagiocephaly or 'flat head syndrome' is a usually temporary deformity of the skull or facial bones caused by prolonged pressure on an area caused by a child lying in the same position for long periods.

611 This narrative is based upon an interview with Heinz Manthey in Lübeck in late November 2016.

612 In German, the title is derogatory in tone, reflecting Elvira's status as a social outsider.

613 S Kent and T Manca (2014), 'A war over mental health professionalism: Scientology versus psychiatry', *Mental Health, Religion & Culture*, 17: pp.1-23

614 S Kent and T Manca (2014), ibid.

615 T Roder, et al (1995), *Psychiatrists: The Men Behind Hitler: The Architects of Horror*, Los Angeles: Freedom Publishing.

616 *Selling Murder: The Killing Films of the Third Reich*, UK (1991), Producers S Lansley, J Mack, written by M Burleigh, Domino Films.

617 This account of the relationship between Elvira Manthey and Ute Hoffman is based upon a November 2016 interview with Hoffman in Bernburg.

GLOSSARY

Aktion – A military-style operation. A term often used by the National Socialist (Nazi) regime to veil governmental crimes, especially in the composition *Sonderaktion* (special campaign) which was used to refer to mass murder.

Aktion T4 – 'Operation T4', the coordinated process of identification of victims suitable for 'euthanasia', mass transportation to killing institutions, their murder, and the issuing of bogus death certificates.

Aktion 14f13 – 'Operation 14f13', the coordinated process of murdering sick concentration camp inmates using former 'euthanasia' killing centres at Bernburg, Pirna-Sonnenstein and Hartheim. The '14f13' code was derived from the record-keeping system for the deaths of concentration camp inmates.

Aktion Reinhard – 'Operation Reinhard', the coordinated process of the mass-murder of Polish Jews in 'extermination camps' (Bełżec, Sobibor, Treblinka and possibly Majdenek). This process involved numerous former *Aktion* T4 staff.

Decentralised euthanasia – The process of killing patients in hospitals by starvation, medication overdose or other means after the cessation of *Aktion* T4 in August 1941.

Gleichschaltung – The process of enforced alignment with Nazi ideology, or 'Nazification', in which the Nazi regime established a system of totalitarian control and coordination over all aspects of society including professions and social institutions.

Heilanstalt – A 'healing' or 'curing' institution such as a sanatorium or psychiatric hospital.

Irrenanstalt – 'insane asylum'– A now outdated and pejorative term for a building used to hold patients experiencing mental illness.

Kinderfachabteilung – 'Special children's wards' – A series of units established as a separate part of hospitals where, under the Nazis, children with disabilities were murdered, usually by lethal injection.

Krankenmorde – The Nazi regime program of mass murder of the sick and disabled, incorporating the separate processes of the murder of psychiatric patients by SS (paramilitary) and Wehrmacht (armed forces) in occupied territories; children's euthanasia; *Aktion* T4; decentralised euthanasia, and sporadic killing of asylum patients in Germany and Austria for bed space.

Lebensunwertes Leben – 'Life unworthy of life' and *Leben ohne Dasein* 'Life without being' were phrases used by advocates of 'euthanasia' to justify the murder of Germans living with disabilities in institutions.

Luminalschema – The procedure for phased lethal dosing of the barbiturate sedative Luminal (phenobarbital) developed by T4 physician Paul Nitsche.

Osteinsatz – A military-style medical operation involving the temporary deployment of *Aktion* T4 medical and nursing staff to the Eastern front for medical support and possibly euthanasia of severely wounded German soldiers and airmen.

Pflegeanstalt – A 'care' institution akin to a nursing home.

Tiergartenstraße 4 – The Berlin address where the adult 'euthanasia' program was headquartered.

Tötungsanstalt – A 'euthanasia' or killing institution, referring to the six *Aktion* T4 killing centres: Brandenburg, Bernburg, Grafeneck, Hadamar, Hartheim, Pirna-Sonnenstein.

Unterwertig – 'mentally inferior'– An outdated pejorative term referring to people with intellectual disabilities.

Volk – The 'people' or the 'nation'. The use of this term in the Nazi era came with connotation of 'race' and was often replaced with the phrase *Volksgemeinschaft*, meaning 'racial community'.

Zwischenanstalt – An intermediate or transitional institution where victims were gathered from different *Heil-* and *Pflegeanstatlen* prior to transportation to killing centres.

APPENDIX I
KEY HISTORICAL FIGURES AND INSTITUTIONS

PEOPLE

Victor Brack – Chief of Office II: Affairs of the Party, State, and the Armed Forces in the *Kanzlei des Führers* (Hitler's Chancellery, the 'KdF'). Brack was assistant to KDF head Phillip Bouhler and was the main functionary in organising *Aktion* T4 and later *Aktion* 14f13 and *Aktion Reinhard*.

Karl Brandt – Medical practitioner, Reich Commissioner for Health and Sanitation, Adolf Hitler's personal physician, and co-leader (with Philipp Bouhler) of *Aktion* T4.

Heinrich Bunke – Medical practitioner, *Aktion* T4 physician, Irmfried Eberl's assistant at Brandenburg and Bernburg 'euthanasia centres' who had assumed the pseudonym 'Dr Rieper'.

Philipp Bouhler – Head of the *Kanzlei des Führers* (Hitler's Chancellery, the 'KdF') and co-leader (with Karl Brandt) of *Aktion* T4.

Leonardo Conti – Medical practitioner, *Reichsgesundheitsführer*, 'Reich Health Leader'.

Irmfried Eberl – Medical practitioner, *Aktion* T4 physician (where he used the alias 'Dr Schneider'), and later Commandant of Treblinka extermination camp.

Ernst Walter Fünfgeld – Medical practitioner, child psychiatrist at paediatric clinic at Magdeburg-Sudenburg, Germany.

Odilo Globočnik – *Schutzstaffel* (Nazi paramilitary, the SS) and Police Leader in the Lublin district of the *General Gouvernment* (German occupation zone) in Poland.

Julius Hallervorden – Medical practitioner, neuropathologist and director of the Neuropathology Department of the Kaiser Wilhelm Institute of *Hirnforschung* (brain research) (KWIHF), part of the *Kaiser-Wilhelm-Gesellschaft zur Förderung der Wissenschaften* (Kaiser Wilhelm Society for the Advancement of Science), an umbrella organisation for leading research institutes in Germany.

Hans Heinze – Medical practitioner, child and adolescent psychiatrist and clinical director of the Görden state psychiatric hospital, Germany.

Werner Heyde – Medical practitioner, professor of psychiatry from Würzburg, Germany, and initial chair of the *Aktion* T4 Medical Committee.

Reinhard Heydrich – Head of *the Sicherheitsdienst* ('SD'; the intelligence and security service, intelligence agency of the SS, and later of the Reich Main Security Office), and Deputy Protector of Bohemia and Moravia (the German occupied part of what is now the Czech Republic).

Appendix 1

Heinrich Himmler – *Reichsführer* in the Nazi regime and head of the *Schutzstaffel* (Nazi paramilitary, the SS).

Adolf Hitler – Führer and Leader of the Nazi regime.

Hermann Paul Nitsche – Medical practitioner, professor of psychiatry at Saxony and Pirna-Sonnenstein, and later chair of *Aktion* T4 Medical Committee.

Franz Stangl – SS *Hauptsturmführer* (Captain). Worked as an upper middle manager in the *Aktion* T4 program and was later Commandant of Sobibor and Treblinka extermination camps.

Hans-Joachim Sewering – Medical practitioner, paediatrician and director of the Schönbrun institution in Dachau, Bavaria, during the Nazi period. He was President of German Medical Association in the 1970s and President-elect of the World Medical Association in 1992, before withdrawing from the position.

Christian Wirth – *Schutzstaffel* (Nazi paramilitary, the SS) *Sturmbannführer* (Major), Inspector of *Aktion* T4 killing centres and *Aktion Reinhard* camps, and Commandant of Bełżec extermination camp.

INSTITUTIONS

BÄK – *Bundesärztekammer* (German Medical Society).

DGPPN – *Deutsche Gesellschaft für Psychiatrie und Psychotherapie, Psychosomatik und Nervenheilkunde* (German Society for Psychiatry, Psychotherapy, Psychosomatics and Neurology).

Einsatzgruppen – 'Special action groups': formations of Nazi paramilitary (*Schutzstaffel or* 'SS'), local and German police, militia and other auxiliaries tasked with the murder of Jews, Communist party officials and other sources of opposition to the Nazi occupation behind lines of advance in Poland and later the USSR. Many massacres of asylum patients were perpetrated by these groups.

Gekrat – *Gemeinnützige Krankentransport* GmbH ('Charitable Ambulance Ltd.') a subdivision of the *Aktion* T4 organization tasked with the transportation of victims to the *Aktion* T4 killing centres.

Gestapo – the *Geheime Staatspolezie* (Secret State Police), abbreviated as 'Gestapo', the official secret police of Nazi Germany.

IMT – International Military Tribunal – The juridical process established at the end of the Second World War by the four victorious Allied powers (Britain, US, USSR and France) for the prosecution of Nazi war crimes. There followed a series of US conducted trials of other perpetrators of crimes in the Nazi regime, including the 'Doctors' trial', that are usually termed the 'Subsequent Nuremberg Proceedings'.

KdF – *Kanzlei des Führers* – Hitler's Chancellery.

KWIHF – Kaiser Wilhelm Institute of *Hirnforschung* (brain research), part of the *Kaiser-Wilhelm-Gesellschaft zur Förderung der Wissenschaften* (Kaiser Wilhelm Society for the Advancement of Science), an umbrella organization for leading research institutes in Germany.

MPG – Max Planck *Gesellschaft* (Max Planck Society), successor organisation of the *Kaiser Wilhelm Gesellschaft* after the Second World War.

The NS or Nazi regime – National Socialist regime – The Nazi government in Germany 1933-1945.

Appendix 1

Reichsgesundheitsführer – Reich Health Leader – Chief of public health.

SA – *Sturmabteilung* (Storm Detachment) – The original paramilitary wing of the Nazi Party (NSDAP) that played a significant role in Adolf Hitler's rise to power in the 1920s and 1930s, usually through political violence. The term 'storm' alludes to the 'storm troops' from the 1914-18 war whose prowess was iconic within the German military. The SA was 'purged' by the SS in 1934, with many of its leadership killed by extrajudicial execution.

SD – *Sicherheitsdienst* ('SD'; the intelligence and security service) – The intelligence agency of the SS.

SS – *Schutzstaffel* (literally Protection Squadron) – A major paramilitary organization of the Nazi party with multiple roles within the Nazi regime; the foremost agency of surveillance and terror within Germany and German-occupied Europe.

Tiergartenstraße 4 – The Berlin address where the adult 'euthanasia' program was headquartered.

APPENDIX 2
OTHER CRIMES PERPETRATED BY NON-GERMAN SCIENTISTS IN THE MID-TWENTIETH CENTURY

To some Germans at the time, the Nuremberg Doctors' trial was an exercise of hypocritical victor's justice. Karl Brandt proclaimed defiantly on the scaffold of the gallows at Landsberg prison on 2 June, 1948: 'How can the nation which holds the lead in human experimentation in any conceivable form, how can that nation dare to accuse and punish other nations which only copied their experimental procedures? And even euthanasia! ...It is, of course, not surprising that the nation which in the face of the history of humanity will forever have to bear the guilt for Hiroshima and Nagasaki, that this nation attempts to hide itself behind moral superlatives'.[1]

As would come to light, Germany was not alone in abusing patients in its institutions of care. Further scandals in medical research would emerge, many that had been perpetrated contemporaneously with, or in some instances before, any of the experiments performed by doctors in Nazi Germany. In 1932, the United States Public Health Service (PHS) embarked on what initially was a six-month prospective study of the effects of untreated syphilis. The research sought to build on the findings of a retrospective study carried out by Norwegian researchers in 1928 that had reported on the clinical and pathological manifestations of untreated syphilis. The American research program—supported by leading US medical associations—continued for 40 years before its determinedly racist application was exposed.

The PHS study of syphilis was based at the Tuskegee Institute in Alabama, around 60 kilometres east of the city of Montgomery. The PHS research team was particularly interested to establish if there were racial determinants in the natural history of the condition. In what is now known as the 'Tuskegee syphilis experiment' around 600 poor African American men from Macon County, Alabama, who had acquired syphilis in their community, were followed up for six months without proper treatment. Promised free care and treatment if they participated, they were subject to toxic and ineffective treatments such as injections of mercury and bismuth. When some research subjects learned of the efficacy of penicillin in treating syphilis, the Tuskegee researchers actively deceived and prevented them from accessing the approved medication, instead providing sugar pills, aspirin or vitamins passed off as penicillin. When any research subject was called up for military service, his untreated condition was used to prevent entry into the military lest a valuable research subject be lost. In some instances, victims of the Tuskegee experiments were subject to painful and unnecessary lumbar punctures after they were misled into believing that the procedure was therapeutic. Despite a national health law in 1943 specifying proven treatment for syphilis, and the 1964 WHO Declaration of Helsinki defining ethical conduct in research, the Tuskegee 'experiment' continued to avoid compliance and due care of its research subjects. When complaints were registered in the mid-1960s, the research program was endorsed by both the American Medical Association and the National Medical Association.

In 1972, a venereologist from the PHS, Dr Peter Buxtun, leaked details of the research program to the American press. The Tuskegee experiment was then halted, but by that time 130 of the men had succumbed to syphilis or its complications, 40 of their wives had been infected, and around 20 of their children were born with congenital syphilis. After a series of hearings, the US Congress passed the *National Research Act* in 1974 to prevent further such Tuskegee-like 'research experimentation'. It was not until May 1997 that then US President Bill Clinton formally apologised for the Tuskegee syphilis experiments.[2]

This was not to be the only outrage involving the syphilis spirochete and American scientists. From 1946 eight PHS investigators in Guatemala deliberately infected more than 1,300 people with syphilis, including many patients in asylums, in order to then study the effectiveness of penicillin against the disease.[3] The researchers enjoyed the full cooperation of local authorities. The history of medical research in the United States is also tarnished with records of experiments infecting prisoners with malaria, or patients in asylums with influenza.[4]

Nor was Nazi Germany the only perpetrator of wartime medical crimes. In what came to be known as 'Unit 731' based in the Pingfang district of Harbin, located in the north east of China, Japanese researchers performed unconsented experiments on more than 250,000 Chinese, Korean and Mongolian civilians and some Anglo-American prisoners of war. The director of Unit 731, Surgeon General Shirō Ishii, had weaponised bubonic plague, which was then used against Chinese civilians. The experiments conducted by Unit 731 between 1937 and 1945 often involved live vivisection after deliberate poisoning with agents used in chemical weapons, or infection with bacteria used in biological weapons.[5] Unlike Karl Brandt and his co-defendants, Ishii was given immunity from prosecution for war crimes by US General Douglas MacArthur, in exchange for information on chemical and biological weapons gathered through the human experimentation conducted by Unit 731. In prosecuting Japanese wartime perpetrators in April 1946, the International Military Tribunal for the Far East (IMTFE) limited the terms of reference related to Unit 731 to experiments with 'poison serums' on Chinese civilians.[6]

APPENDIX 2 ENDNOTES

1 U Schmidt (2007), *Karl Brandt: The Nazi Doctor*, London: Hambledon Continuum, p.381

2 See J Jones (1981), *Bad Blood: The Tuskegee Syphilis Experiment*, New York: Free Press; and S Reverby (2009), *Examining Tuskegee: The Infamous Syphilis Study and its Legacy*, Chapel Hill: University of North Carolina Press.

3 *Ethically Impossible: STD Research in Guatemala from 1946 to 1948*, Presidential Commission for the Study of Bioethical Issues, published 2011: http://bioethics.gov/sites/default/files/Ethically%20Impossible%20%28with%20linked%20historical%20documents%29%202.7.13.pdf

4 See B *Brody (1998)*, *The Ethics of Biomedical Research: An international perspective*, New York: Oxford University Press.

5 J Guillemin (2017), *Hidden Atrocities: Japanese Germ Warfare and American Obstruction of Justice at the Tokyo Trial*, New York: Columbia University Press.

6 See H Gold (2006), *Unit 731: Testimony*, Boston: Tuttle Publishing.

APPENDIX 3
NOTE ON RESEARCH AND SOURCES

The account of Elvira Hempel-Manthey's life described in this book is drawn from several sources, predominantly from those written and spoken by Elvira herself. Unless otherwise indicated, details about her experiences were elicited from her self-published book, *Die Hempelsche - Das Schicksal eines deutschen Kindes, das 1940 vor der Gaskammer umkehren durfte*. The authors also referred to a published English translation, *Elvira – The fate of a German child who cheated death in the Gas Chamber in 1940*, with further research to clarify aspects of the book undertaken by translator Antje Hammond. A copy of the book was sourced from Elvira's widower, Heinz Manthey, who assisted with its original publication and its subsequent productions and dissemination.

In late 2016 Herr Manthey met with the one of the authors, Michael Robertson, and translator Antje Hammond. During this meeting Herr Manthey clarified and expanded on aspects of Elvira's book, provided copies of illustrations and records, and shared reflections on his life with Elvira and in the years since her death.

In 2015 and 2016 the authors met with historian Dr Ute Hoffman, head of the Memorial for Victims of Nazi 'Euthanasia', Bernburg.

Elvira gave her original Uchtspringe file to Dr Hoffman, who was able to further explain aspects of these documents as well as her association with Elvira. The authors also utilised accounts of Elvira's experiences in the oral testimonies she gave in the early 2000s, including to the Russell Tribunal on Human Rights in Psychiatry, and the United States Holocaust Memorial Museum:

> E Manthey (1995), *Die Hempelsche - Das Schicksal eines deutschen Kindes, das 1940 vor der Gaskammer umkehren durfte*, Lübeck: Hempel-Verlag Heinz Manthey (German and English editions)
>
> 'Testimony of Elvira Manthey (born Hempel) at The Russell Tribunal on Human Rights in Psychiatry, June 30, 2001', Berlin: Germany (English translation) http://www.freedom-of-thought.de/rt/manthey.htm [Accessed 5 February 2019]
>
> 'Oral history interview with Elvira Manthey', January 13, 2003, United States Holocaust Memorial Museum (interviewer S Stept), collection accession number: 2003.485.8, RG number: RG-50.718.0008, https://collections.ushmm.org/search/catalog/irn60516 [Accessed 5 February 2019].
>
> See also 'Oral History: Benno Müller-Hill, Antje Kosemund, Paul Eggert, and Elvira Manthey describe the Euthanasia Program', https://encyclopedia.ushmm.org/content/en/oral-history/benno-mueller-hill-antje-kosemund-paul-eggert-and-elvira-manthey-describe-the-euthanasia-program

In broader research, the authors have drawn upon their professional experience in history, psychiatry and bioethics, including research focused on the *Krankenmorde*. This has included fieldwork at the following museums, memorials, archives, and other institutional sites:

Appendix 3

museums/memorials/archives at the six 'euthanasia' killing centre sites in Germany and Austria: Brandenburg (where author Astrid Ley is Director), Bernburg, Pirna-Sonnenstein, Grafenek, Hadamar, and Hartheim.

museums/memorials/archives at hospitals in Germany, Austria, Poland: Christophsbad hospital, Goppingen, Germany; Eglfing-Haar hospital, near Munich, Germany; Görden hospital, Brandenburg an der Havel, Germany; Miedzyrzecz (Obrzyce) psychiatric hospital, Poland; Otto Wagner and the Steinhof Psychiatric Hospital, Vienna, Austria.

archives: Archiv des Bezirk Oberbayern, Munich, Germany; Bundesarchiv, Abt. Militärarchiv, Freiburg, Germany.

museums/memorials at concentration and extermination camp sites: Sachsenhausen in Germany; Treblinka, Bełżec, Majdanek in Poland; and Terezin in the Czech Republic.

other institutions/programs: Memorium Nuremberg, 'Medicine without Conscience. The Nuremberg Doctors' Trial' program, and Human Rights Office of the City of Nuremberg human rights program, Germany; Sächsisches Psychiatriemuseum, Leipzig, Germany; Yad Vashem, Museum and Archives, Jerusalem, Israel.

We thank and acknowledge the people working at these sites—historians, curators, educators, ethicists, physicians, advocates—who generously shared their time and expertise with us, in particular those mentioned in our Acknowledgements section.

APPENDIX 4
SELECT ENGLISH LANGUAGE BIBLIOGRAPHY

In this section we have highlighted some of the English language resources (including translations of German texts where available) which will for the interested reader both complement and add depth to the focus areas mentioned.

COMPREHENSIVE HISTORIES OF NAZI 'EUTHANASIA' AND MEDICAL CRIMES UNDER NAZISM

S Benedict and L Shields (eds), *Nurses and Midwives in Nazi Germany: The 'Euthanasia Programs'* (Routledge Studies in Modern European History) 1st Edition, New York: Routledge, 2014.

M Burleigh (1997), *Ethics and Extermination – Reflections on Nazi Genocide*. New York: Cambridge.

— (2002 a), *Death and Deliverance - Euthanasia in Germany 1900-1945*, London: Pan.

— (2002 b), 'The Legacy of Nazi Medicine in Context', in F Nicosia and J Huener (eds), *Medicine and Medical Ethics in Nazi Germany*, New York: Berghahn Books, pp.112–127

A Caplan (1992), 'The Doctors' trial and Analogies to the Holocaust in Contemporary Bioethical Debates', in G Annas and M Grodin (eds), *The Nazi Doctors and the Nuremberg Code*, New York: Oxford University Press, pp.258–275

H Czech (2014), 'Abusive medical practices on 'euthanasia' victims in Austria during and after World War II', in S Rubenfeld S, S Benedict (eds), *Human subjects research after the Holocaust*, Cham: Springe International Publishing, pp.109–126

H Friedlander (1995), *The Origins of Nazi Genocide: From Euthanasia to the Final Solution*, Chapel Hill: University of North Carolina Press.

— (2002), 'Physicians as Killers in Nazi Germany', in F Nicosia and J Huener (eds), ibid, pp.59–76

M Kater (1989), *Doctors under Hitler*, Chapel Hill: University of North Carolina Press.

N Kulish and S Mekhennet (2014), *The Eternal Nazi: From Mauthausen to Cairo, the Relentless Pursuit of SS Doctor Aribert Heim*, New York: Vintage.

A Ley and A Hinz-Wessels (2012), *The 'Euthanasia Institution' of Brandenburg an der Havel*, Berlin: Metropol.

R Lifton (1986), *The Nazi Doctors*, New York: Basic Books.

B Müller-Hill (1998), *Murderous Science: Elimination by Scientific Selection of Jews, Gypsies, and Others in Germany, 1933-1945*, G Fraser (trans), Plainview, NY: Cold Spring Harbor Laboratory Press.

R Pommerin (1979), *Sterilization of the Rhineland Bastards. The Fate of a Colored German Minority from 1918 to 1937*, Dusseldorf: Droste.

U Schmidt (2004), *Justice at Nuremberg*, London: Palgrave Macmillan.

— (2007), *Karl Brandt: The Nazi Doctor*, London: Hambledon Continuum.

H Schmuhl (2009), 'Brain Research and the Murder of the Sick: The Kaiser Wilhelm Institute for Brain Research 1937-1945', in S Heim, C Sachse and M Walker (eds), *The Kaiser Wilhelm Society under National Socialism*, New York: Cambridge University Press, pp.99–119

W Seidelman (2002), 'Pathology of Memory: German Medical Science and the Crimes of the Third Reich', in F Nicosia and J Huener (eds) ibid, pp.93–111

— (2014). "Requiescat sine Pace': Recollections and Reflections on the World Medical Association, the Case of Prof. Dr Hans Joachim Sewering and the Murder of Babette Fröwis', in V Roelcke, S Topp and E Lepicard (eds), *Silence, Scapegoats, Self-Reflection - The Shadow of Nazi Medicine and Bioethics*, Gottingen: V&R Unipress, pp.281–300

GENERAL AND CULTURAL HISTORIES OF NAZISM

J Bendersky (2007), *A Concise History of Nazi Germany*, Lanham MA: Rowman and Littlefield.

M Burleigh (2000), *The Third Reich - a New History*, London: MacMillan.

M Dederichs (2009), *Heydrich: The Face of Evil*, Drexel Hill PA: Casemate.

R Evans (2006), *The Third Reich in Power 1933-1939*, London: Penguin.

R Gerwarth (2011), *Hitler's Hangman: The Life of Heydrich*, London: Yale University Press.

L Goldensohn (2004), *The Nuremburg Interviews*, New York: Vintage.

B Hinz (1979), *Art in the Third Reich*, New York: Random House.

— (2000), *'Degenerate Art' - Art in the Third Reich*, New York: Random House.

H Höhne (1969), *The Order of the Death's Head: The Story of Hitler's SS*, London: Penguin.

I Kershaw (2000), *The Nazi Dictatorship: Problems and Perspectives of Interpretation* 4th Edn, New York: Oxford University Press.

V Klemperer (1998), *I Shall Bear Witness: The Diaries of Victor Klemperer, 1933-41*, M Chalmers (trans), London: Weidenfeld & Nicolson.

C Koonz (2002), *The Nazi Conscience*, Cambridge, Mass: Belknap.

G Sereny (1983), *Into That Darkness: An Examination of Conscience*, New York: Vintage.

— (2002), *The Healing Wound - Reflections on Germany 1938-2001*, New York: Norton.

W Shirer (1941), *Berlin Diary: The Journal of a Foreign Correspondent 1934–1941*, New York: Knopf.

N Stargardt (2015), *The German War*, London: The Bodley Head.

A Weale (2010), *The SS - a New History*, London: Little Brown.

RESISTANCE TO THE NAZI REGIME

A Dumbach and J Newborn (1986), *Sophie Scholl and the White Rose*, London: One World.

A Gill (1994), *An Honourable Defeat; a History of the German Resistance to Hitler*, London: Heinemann.

E Griech-Polell (2002), *Bishop Von Galen: German Catholicism and National Socialism*, New Haven CT: Yale.

R Krieg (2004), *Catholic Theologians in Nazi Germany*, New York: Continuum.

C von Galen (2006), *Bishop Clemens August Graf Von Galen - Files, Letters and Sermons from 1933 to 1946* 2nd Edition, Paderborn: Ferdinand Schöningh.

I Scholl (1983), *The White Rose: Munich, 1942–1943*, A Schultz (trans), Middletown, CT: Wesleyan University Press.

HISTORIES AND ANALYSES OF THE HOLOCAUST

Y Arad (1987), *Belzec, Sobibor, Treblinka: The Operation Reinhard Death Camps*, Bloomington, Il: Indiana University Press.

Y Bauer (2001), *Rethinking the Holocaust*, London: Yale University Press.

I Baxter (2016), *The SS of Treblinka*, Stroud: The History Press.

D Bloxham (2009), *The Final Solution: A Genocide*, New York: Oxford University Press.

M Broszat (1981), *The Hitler State: The Foundation and Development of the Internal Structure of the Third Reich*, London: Longman.

C Browning (1992), *The path to genocide: essays on launching the final solution*, Cambridge: Cambridge University Press.

— (2004), *The Origins of the Final Solution - the Evolution of Nazi Jewish Policy, September 1939-March 1942*, Lincoln: University of Nebraska Press.

F Carr (2003), *Germany's Black Holocaust, 1890-1945*, Kearney NE: Morris Publishing.

Z Garber and B Zuckerman (2004), 'Double Takes: Thinking and Rethinking Issues of Modern Judaism in Ancient Contexts', *Studies in the Shoah Vol. 26*, Lanham, MD: University Press of America.

M Gilbert (2006), *Kristallnacht: Prelude to Destruction*, New York: Harper Collins.

D Goldhagen (1996), *Hitler's Willing Executioners: Ordinary Germans and the Holocaust*, New York: Knopf.

H Friedlander (2000), 'Step by Step: The Expansion of Murder, 1939-1941', in O Bartov (ed), *The Holocaust - Origins, Implementation, Aftermath*, London: Routledge, pp.63–76

S Friedländer (2007), *The Years of Extermination: Nazi Germany and the Jews 1939-1945*, New York: Harper Collins.

R Hilberg (1961), *The Destruction of the European Jews*, London: W H Allen.

E Klee, W Dressen and V Riess (2002), *The Good Old Days: The*

Holocaust as Seen by Its Perpetrators and Bystanders, D Burnstone (trans), Old Saybrook: Konecky & Konecky.

P Longerich (2010), *Holocaust: The Nazi Persecution and Murder of the Jews*, London: Oxford University Press.

R MacNair (2005), *Perpetration-Induced Traumatic Stress: The Psychological Consequences of Killing*, New York: Authors Choice Press.

T Mason (1995), 'Intention and explanation A current controversy about the interpretation of National Socialism', in J Caplan (ed), *Nazism, Fascism and the Working Class*, Cambridge: Cambridge University Press, pp.212–230

K Schleunes (1970), *The Twisted Road to Auschwitz; Nazi Policy Toward German Jews, 1933–1939*, Urbana: University of Illinois Press.

GENOCIDE STUDIES

R Lemkin (1944), 'Chapter ix: 'Genocide'- a New Term and New Conception for the Destruction of Nations', in *Axis Rule in Occupied Europe - Laws of Occupation – Analysis of Government – Proposals for Redress*, Washington DC: Carnegie Endowment for International Peace, pp.79–95

M Shaw (2015), *What is Genocide?* Cambridge: Polity.

T Snyder (2010), *Bloodlands: Europe between Hitler and Stalin*, New York: Basic Books.

C Tatz and W Higgins (2016), *The Magnitude of Genocide*, Santa Barbara: Praeger, 2016.

C Tatz (2005), 'The doctorhood of genocide', in *Genocide and Human Rights: A Philosophical Guide*, Roth, JK (ed). Hampshire: Palgrave Macmillan, pp.82–94

— (2017 a), *Australia's Unthinkable Genocide*, Bloomington: Xlibris.

— (2017 b), 'Too near and too far: Australia's reluctance to name and prosecute genocide', in N Marczak and K Shields (eds), *Genocide Perspectives V - A Global Crime, Australian Voices*, UTS ePRESS, pp.51–68

GERMAN POST-WAR ENGAGEMENT WITH THE NAZI PERIOD

T Adorno (2005), 'The Meaning of Working Through the Past', *Critical Models: Interventions and Catchwords*, H Pickford (trans), New York: Columbia University Press, pp.89–103

P Baldwin (1990), *Hitler, the Holocaust and the Historians Debate*, Boston: Beacon Press.

M Bryant (2005), *Confronting the 'Good Death': Nazi 'Euthanasia' on Trial 1945–1953*, Boulder: University Press of Colorado.

— (2014), *Eyewitness to Genocide – The Operation Reinhart Death Camp Trials 1955-66*, Knoxville. University of Tennessee Press.

C Cocks (1997), *Psychotherapy in the Third Reich: The Göring Institute*, New York: Oxford.

F Fischer (1986), *From Kaiserreich to Third Reich: Elements of Continuity in German History 1871-1945*, London: Allen & Unwin.

J Goggin and E Brockman-Goggin (2001), *Death of a Jewish Science: Psychoanalysis in the Third Reich*, West Lafayette: Purdue University Press.

S Kattago (2001), *Ambiguous Memory: The Nazi Past and German National Identity*, Westport: Praeger.

S Knittel (2016), 'Autobiograpy, Moral Witnessing and the Disturbing Memory of Nazi Euthanasia', in M Fulbrook, S Wienand, J Wagner (eds), *Reverberations of Nazi Violence in Germany and Beyond*, pp.65-84

D Lipstadt (1993), *Denying the Holocaust: The Growing Assault on Truth and Memory*. New York: Plume.

A Taylor (1945), *The Course of German History*, London: Hamish Hamilton.

POST-WAR COMMEMORATION AND MEMORIALISATION OF VICTIMS OF NAZISM

A Alba (2015), *The Holocaust Memorial Museum - Sacred Secular Space*, Basingstoke: Palgrave.

I Buruma (1994), *The wages of guilt - memories of war in Germany and Japan*, New York: New York Review of Books.

E Claasen (1987[1969]), *Ich, die Steri*, Bonn: Psychiatrie-Verlag.

R Dome (2008), *Nebel im August*, Munich: Random House.

F Fischer (1986), *From Kaiserreich to Third Reich: Elements of Continuity in German History 1871-1945*, London: Allen & Unwin.

N Frei (2002), *Adenauer's Germany and the Nazi Past: The Politics of Amnesty and Integration*, New York: Columbia University Press.

A Kaufmann (2007 [1986]), *Totenwagen. Kindheit am Spiegelgrund*, Mechthild Podzeit-Lütjen (ed), Vienna: Mandelbaum.

S Knittel (2010), 'Remembering Euthanasia: Grafeneck in the Past, Present and Future', in B Niven and C Paver (eds), *Memorialization in Germany since 1945*, Houndmills: Palgrave, pp.124–133

D Magilow and L Silverman (2015), *Holocaust Representations in History: An Introduction*, New York: Bloomsbury.

A Mitscherlich and M Mitscherlich (1975), *The Inability to Mourn. Principles of Collective Behavior*, New York: Grove Press.

C Pross (1998), *Paying for the Past - The Struggle over Reparations for Surviving Victims of the Nazi Terror*, Baltimore: Johns Hopkins University Press.

G Rosenfeld (2015), *Hi Hitler! How the Nazi Past Is Being Normalized in Contemporary Culture*, London: Cambridge University Press.

G Schmid-Krebs and R Brüggemann (2007), *Verortungen der Seele - Psychiatrie-Museen und verwandte Einrichtungen in Europa ('Locations of the soul - Psychiatry museums and related institutions in Europe')*, Goppingen: Mabuse-Verlag.

R Surmannn (2014), 'Rehabilitation and Indemnification for the Victims of Forced Sterilization and 'Euthanasia'. The West German Policies of 'Compensation' ('Wiedergutmachung')', in V Roelcke, S Topp, E Lepicard (eds), *Silence, Scapegoats, Self-reflection: The Shadow of Nazi Medical Crimes on Medicine and Bioethics*, pp.113–130

D Vyleta (2011), *The Quiet Twin*, London: Bloomsbury.

EARLY TWENTIETH CENTURY AND CONTEMPORARY EUGENICS

A Bashford & P Levine (2010), *The Oxford Handbook of the History of Eugenics*, New York: Oxford University Press.

E Black (2003), *War against the Weak: Eugenics and America's Campaign to Create a Master Race*, New York: Basic Books.

A Cohen (2016), *Imbeciles: The Supreme Court, American Eugenics, and the Sterilization of Carrie Buck*, New York: Penguin.

R Hernstein and C Murray (1994), *The Bell Curve: Intelligence and Class Structure in American Life*, New York: The Free Press.

J Huxley (1957), 'Transhumanism', in *New Bottles for New Wine* London: Chatto & Windus, pp.13–17

R Kluchin (2009), *Fit to Be Tied: Sterilization and Reproductive Rights in America 1950–1980*, New Brunswick: Rutgers University Press.

S Kühl (1994), *The Nazi Connection - Eugenics, American Racism, and German National Socialism*, New York: Oxford University Press.

C Mills (2015), 'Liberal Eugenics, Human Enhancement and the Concept of the Normal' in D Meacham (ed) *Medicine and Society, New Perspectives in Continental Philosophy*, Dordrecht: Springer, p.179–194

— (2017), 'Biopolitics and human reproduction', in S Prozorov and S Rentea (eds), *The Routledge Handbook of Biopolitics*, Abingdon: Routledge, pp.281–294

A Stern (2005), *Eugenic Nation: Faults and Frontiers of Better Breeding in Modern America*, Berkeley: University of California Press.

L Zenderland (1998), *Measuring Minds: Henry Herbert Goddard and the origins of American intelligence testing*, New York: Cambridge University Press.

HISTORY OF PSYCHIATRY

A Deutsch (1948), *The Shame of the States*, New York: Harcourt, Brace.

D Healy (1997), *The Antidepressant Era*, Cambridge: Harvard University Press.

H Jones (2013), *Byberry State Hospital*, Charleston: Arcadia

D Schreber (1903/2000), *Memories of My Nervous Illness*, I Macalpine and R Hunter (trans), New York: New York Review of Books.

E Shorter (1997), *A History of Psychiatry: From the Era of the Asylum to the Age of Prozac*, New York: Wiley.

S Taylor (2009), *Acts of Conscience: World War II, Mental Institutions, and Religious Objectors* Syracuse, NY: Syracuse University Press.

R Whitaker (2002), *Mad in America: Bad Science, Bad Medicine, and the Enduring Mistreatment of the Mentally Ill*, New York: Perseus.

BIOETHICS

B Brody (1998), *The Ethics of Biomedical Research: An international perspective*, New York: Oxford University Press

T Chelouche and G Brahmer (2013), *Casebook on Bioethics and the Holocaust*, Haifa: UNESCO Chair in Bioethics.

J Keown (2002), *Euthanasia, Ethics and Public Policy: An Argument Against Legalisation*, London: Cambridge University Press.

I Kerridge and C Stewart (2009), *Ethics and law for the health professions*, Sydney: The Federation Press.

J Rachels (1996), *The End of Life*, Oxford: Oxford University Press, pp.175–178

P Singer (1993), *Practical Ethics (2nd Edition)*, Cambridge: Cambridge University Press.

S Stefan (2016), *Rational Suicide, Irrational Laws: Examining Current Approaches to Suicide in Policy and Law*, New York: Oxford University Press.

BIOPOLITICS AND BIOPOWER

G Agamben (1988), *Homo Sacer: Sovereign Power and Bare Life*, Stanford: Stanford University Press.

Z Bauman (1989), *Modernity and the Holocaust*, Ithaca, N.Y: Cornell University Press.

A Brandt (2007), *The Cigarette Century: The Rise, Fall, and Deadly Persistence of the Product that Defined America*, New York: Basic Books.

M Foucault (1969), *The Archaeology of Knowledge*, A Sheridan (trans), New York: Pantheon Books.

— (1981), *The History of Sexuality Volume I - an Introduction*, London: Harmondsworth: Penguin.

— (2004), *The Birth of Biopolitics - Lectures at the College De France 1978-9*, G Burchell (trans), New York: Picador.

C Mayes (2016), *The Biopolitics of Lifestyle*, Oxford: Routledge.

R Proctor (1999), *The Nazi War on Cancer*, Princeton: Princeton University Press.

DISABILITY RIGHTS

D Braddock and S Parish (2002), 'An Institutional History of Disability', in D Braddock (ed), *Disability at the Dawn of the 21st Century and The State of the States*, Washington: American Association Intellectual Disabilities, pp.11–68

F Campbell (2010), *Contours of Ableism: The Production of Disability and Abledness*, Melbourne: Palgrave Macmillan.

M Dudley, D Silove and F Gale (2012), *Mental health and human rights: vision, praxis, and courage*, Oxford UK: Oxford University Press.

D Mitchell and S Snyder (2015), *The Biopolitics of Disability: Neoliberalism, Ablenationalism, and Peripheral Embodiment*, Ann Arbor: University of Michigan Press.

M Oliver (1983), *Social Work with Disabled People*, Basingstoke: Macmillan.

— (1990), *The Politics of Disablement*, Basingstoke: Macmillan.

— (1996), *Understanding disability: from theory to practice*, Basingstoke: Macmillan.

J Scully (2008), *Disability bioethics: Moral bodies, Moral difference*. Gloucester: Rowman & Littlefield.

Appendix 4

T Shakespeare (2006), *Disability rights and wrongs*, London: Routledge.

L Staniland (2012), *Public Perceptions of Disabled People: Evidence from the British Social Attitudes Survey 2009*, London: Office for Disability Issues.

AUTHOR BIOGRAPHIES

MICHAEL ROBERTSON

Michael Robertson is a consultant psychiatrist, Clinical Associate Professor of Mental Health Ethics (Sydney Health Ethics centre, University of Sydney) and a Visiting Professorial Fellow at the Sydney Jewish Museum. He has researched, taught and written extensively on psychiatric ethics, involuntary psychiatric treatment, psychotherapy, psychological trauma and the psychiatric profession under National Socialism.

ASTRID LEY

Astrid Ley, PhD, is a historian and historian of medicine. She is curator of the Brandenburg T4 memorial, and deputy director and academic head of the Sachsenhausen and Brandenburg memorials' foundation. Her main research interest is medicine under National Socialism. At Sachsenhausen, she curated several permanent exhibitions, including *The Euthanasia Institution at Brandenburg an der Havel*. She has published numerous articles in academic periodicals and edited two anthologies. She is a lecturer at Berlin Free University.

EDWINA LIGHT

Edwina Light is a postdoctoral research fellow at Sydney Health Ethics (SHE) at the University of Sydney, and a visiting fellow at the Sydney Jewish Museum. With a background in health journalism and communications, she completed her bioethics training at SHE. Her research seeks to inform public debate and policymaking where matters of health ethics, policy and law intersect, with a particular focus on mental health ethics. Her research publications include works on mental health policy, involuntary treatment and systems of coercion, the experiences and perspectives of stakeholders, and the psychiatric profession under National Socialism.

www.ingramcontent.com/pod-product-compliance
Lightning Source LLC
Chambersburg PA
CBHW050927240426
43670CB00023B/2951